STATISTICAL INFERENCE AND PROBABILITY

THE SAGE QUANTITATIVE RESEARCH KIT

Statistical Inference and Probability by *John MacInnes* is the 3rd volume in *The SAGE Quantitative Research Kit*. This book can be used together with the other titles in the *Kit* as a comprehensive guide to the process of doing quantitative research, but is equally valuable on its own as a practical introduction to inferential statistics and probability.

Editors of The SAGE Quantitative Research Kit:

Malcolm Williams – *Cardiff University, UK*

Richard D. Wiggins – *UCL Social Research Institute, UK*

D. Betsy McCoach – *University of Connecticut, USA*

Founding editor:

The late W. Paul Vogt – *Illinois State University, USA*

STATISTICAL INFERENCE AND PROBABILITY

JOHN MACINNES

Los Angeles | London | New Delhi
Singapore | Washington DC | Melbourne

THE SAGE QUANTITATIVE RESEARCH KIT

Los Angeles | London | New Delhi
Singapore | Washington DC | Melbourne

SAGE Publications Ltd
1 Oliver's Yard
55 City Road
London EC1Y 1SP

SAGE Publications Inc.
2455 Teller Road
Thousand Oaks, California 91320

SAGE Publications India Pvt Ltd
B 1/I 1 Mohan Cooperative Industrial Area
Mathura Road
New Delhi 110 044

SAGE Publications Asia-Pacific Pte Ltd
3 Church Street
#10-04 Samsung Hub
Singapore 049483

Editor: Jai Seaman
Assistant editor: Charlotte Bush
Production editor: Manmeet Kaur Tura
Copyeditor: QuADS Prepress Pvt Ltd
Proofreader: Derek Markham
Indexer: Caroline Eley
Marketing manager: Susheel Gokarakonda
Cover design: Shaun Mercier
Typeset by: C&M Digitals (P) Ltd, Chennai, India

© John MacInnes 2022

This volume published as part of *The SAGE Quantitative Research Kit* (2021), edited by Malcolm Williams, Richard D. Wiggins and D. Betsy McCoach.

Library of Congress Control Number: 2020945819

British Library Cataloguing in Publication data

A catalogue record for this book is available from the British Library

ISBN 978-1-5264-2416-7

For Esther

CONTENTS

LIST OF FIGURES AND TABLES

List of figures

List of tables

ABOUT THE AUTHOR

John MacInnes is emeritus professor of sociology and statistics at the University of Edinburgh, a Fellow of the Academy of Social Sciences and a Chartered Statistician. He was Vice President of the Royal Statistical Society (2019–2020), Strategic Advisor to the ESRC and British Academy on quantitative skills development and helped develop the UK Q-Step programme. His substantive research interests have ranged from social demography to gender studies, nationalism and employment relations, confirming John Tukey's claim that statisticians get to play in everyone's backyard.

1

THE CHALLENGE AND PROMISE OF INFERENCE

Chapter Overview

What is inference?

Inference is the process of drawing reasoned but risky conclusions from empirical evidence.

I enter a room and see a person with a smoking gun in their hand beside a body with gunshot wounds. I could *infer* that the person holding the gun had shot the other one. Other evidence might be relevant to my conclusion. I might know the two people were mortal enemies. I might just have heard a gunshot. However, my inference and the conclusion I draw from it would be a risky one because there is only some probability that it is correct. Perhaps the body is a suicide, and the person now holding the gun had been desperately trying to wrestle it from the victim. I did not witness the shot directly. Even if I had, I would still need to be sure that it was not some visual trick or illusion and that everything was indeed as it seemed on the surface.

This is the situation we face with most evidence. Many of the processes we try to understand are invisible. We cannot 'see' class, ethnic discrimination, economic growth or the rise of populism directly – if we could, there would be little need for social science – rather, we can collect evidence about the results of these processes and build models of what we think may be happening to produce that evidence. That is what scientific inference comprises.

It is helpful to think of three categories of inference: (1) *informal* inference, which is something intuitive we do all the time; (2) *scientific* inference, which is a set of rules laid down to minimise the risk of drawing unsound conclusions; and (3) *statistical* inference, which is the part of scientific inference that deals with generalising evidence taken from samples to the whole populations from which these samples have been drawn. Almost all the evidence we ever work with is a sample of some kind, so that statistical inference is a fundamental part of the scientific method.

Informal inference: the tyranny of causal narratives

We are inference machines who understand the world around us by constantly drawing barely conscious conclusions, and effortlessly constructing plausible causal narratives based upon them, both to justify ourselves to others and to reassure ourselves that we understand the world and our place in it. Much of our experience of the world proceeds by induction, whereby finding repeated examples of the same thing or process leads us to expect that under similar conditions we will nearly always find them again. Without the existential reassurance provided by induction, the world would appear as a rather terrifying and unpredictable chaos. Our everyday behaviour in the world is rooted in continually updating our awareness of what is happening around us, drawing conclusions from it and telling stories to ourselves. I was happy today *because* it was sunny. I got better *because* I took that medicine. The wood in the stove burned *because* I set fire to it. I was late for work *because* the train was delayed. The object fell *because* of gravity. My accident happened *because* the cyclist didn't see me coming. We usually process these causal stories below the level of conscious awareness or calculation. I recognise that person *because* I have seen them before. I know that person is angry *because* of their facial expression and so on. I stepped out onto the road to cross it *because* I didn't hear any traffic (and consequently collided with the cyclist I didn't look out for).

Most of these inferences will be at least accurate enough to ensure our well-being. We learn to judge the speed and likely behaviour of traffic, the amount of effort we need to put into writing an essay that will pass or the kind of clothes that make us warm, comfortable or feel sexy. We are all extremely good at making causal associations: links between one phenomenon or pattern and another. When we look at someone's face, without even having to think consciously about it, we can infer how they are feeling, whether they are angry, happy, sad or bored. We make an association between their facial expression, the configuration of their eyes and eyebrows, mouth, nose and forehead and how we imagine them to be feeling. We will also, usually without consciously thinking about it, produce an explanation of the feeling we suppose them to be experiencing.

However, many of our informal inferences may be wildly wrong and our causal narratives usually flatter to deceive. I do not know that the medicine cured my ailment: perhaps I would have recovered without it. Perhaps if I had taken the bus I would have arrived on time, or any number of other events might have occurred. I could have left the house earlier and caught a different train. Maybe it was the sunshine that made me happy, but how could I know? Perhaps my mood would have been the same had the day been cloudy and overcast. Until Newton's work on motion, I might have said that the object fell because it was heavy. In the early 18th century, I might have said the wood burned because it contained phlogiston. For me to have that accident a whole set of circumstances were required – from the invention of the bicycle to my decision to cross the road, the existence of the road, all the reasons for the cyclist to be there too at that precise time, their inability to take evasive action, the weather conditions at the time and so on and on and on.

Thus, while we might imagine that we understand the world around us through causal narratives, our perceptions have at best a tenuous link to empirical reality, and any but the most superficial understanding would in principle require virtual omniscience about all manner of diverse causal chains and their history. It would be impossible. Worse, it would leave no room for any knowledge of the world to be of any use! In his *Philosophical Essay on Probabilities*, published in 1814, the French mathematician Pierre-Simon Laplace imagined an omniscient 'intelligence' that later philosophers came to refer to as his 'demon':

> We may regard the present state of the universe as the effect of its past and the cause of its future. An intellect which at a certain moment would know all forces that set nature in motion, and all positions of all items of which nature is composed, if this intellect were also vast enough to submit these data to analysis, it would embrace in a single formula the movements of the greatest bodies of the universe and those of the tiniest atom; for such an intellect nothing would be uncertain and the future just like the past would be present before its eyes. (Laplace, 1951, p. 4)

This vision of a purely causal world without randomness later came to be known as a 'block universe': a world in which determinism squeezed out any possibility of change or evolution. In such a world, knowing everything would, paradoxically, be utterly valueless, since the fact of that knowledge could not change anything in the present, past or future. To have room to breathe, and space for change, we *need* randomness and probability.

Cognitive illusions

Cognitive psychologists have a fairly good picture of the many kinds of systematic biases that drive our everyday inferences. Kahneman (2011) calls them 'cognitive illusions', which are similar to the visual illusions or tricks of perspective that you may be familiar with, in which our vision tricks us into seeing things that are not really there. We usually pay too much attention to easily available evidence, regardless of its relevance or quality, a phenomenon he refers to as 'WYSIATI: What You See Is All There Is'. We may also substitute questions we find hard to answer with similar ones that we find easier, but which may have little to do with the issue at hand. 'Is that person fair and reasonable?' may morph into 'do I like them?' We avoid numerical calculations (they demand effort, are slow and prone to mistakes) and substitute rough estimates which we then treat with too much certainty. We see patterns where there are none. We rarely evaluate or check up on our inferences or predictions and may even persuade ourselves that we never actually held a belief that later turned out to be faulty. Above all, we practice confirmation bias remorselessly. We avoid or disregard evidence that goes against our existing beliefs, while seizing upon information that can be interpreted to support them. We can persist in this even in the face of overwhelming odds, until something truly traumatic or catastrophic forces a change in our views. Kahneman has therefore described our informal inference making as a 'machine for jumping to conclusions'. It is plausible (but not more than that) to imagine that such a machine had evolutionary survival value. Imagining a potential threat that turns out to be a false alarm carries little penalty. Ignoring one that turns out to be real might be deadly.

These processes are one reason why the history of civilisation is adorned with all manner of fanciful tales and characters such as werewolves, demons, spirits, miracles and so on. The range of bizarre beliefs that credulous societies have solemnly subscribed to is long. Until a few hundred years ago, inferring from the pattern of sunrise and sunset that the sun circled the earth, which was flat, and that matter was solid, would simply have appeared to be obvious. Do not, for a moment, believe that the society we inhabit today is an exception. No society has ever believed itself to be systematically deluded. Because of our ability to effortlessly and systematically deceive ourselves, we need to turn off our semi-automatic cognitive machinery in order to think scientifically. It is probability that allows us to do this, by imagining the world not in terms of causal chains (although these may well exist) but in terms of collections of events with different probabilities of occurring. We may observe patterns of association within these events that are themselves probabilistic in character, so that we can explore what events may increase or decrease the probability of other events taking place. However, most important of all, we grant this external world of events the authority to test our theories about the world, leaving nothing to the reputation or skill of the scientific investigator, no matter how renowned. Scientific inference puts empirical evidence in charge. As we shall see later, this has the paradoxical effect of changing the nature of discovery, knowledge and insight. Rather than an ever-growing accumulation of proven 'facts', discovery mostly comprises the revision or destruction of parts of the provisional knowledge we currently hold, as we become more aware of just how boundless is our ignorance. It becomes, to borrow a phrase, the discovery of hitherto unknown unknowns.

Scientific inference

Scientific inference attempts to circumvent these cognitive biases by insisting that inference is based on evidence marshalled in such a way that the individual scientist has as little control over it as possible, in order to maximise the probability that the conclusions reached are sound by tackling confirmation bias, tunnel vision, wobbly logic or loaded arguments. Paradoxically, taking the scientist out of the science often requires complex research designs that require a good deal of scientific expertise and experience to construct. This does not mean that 'the facts speak for themselves'. If facts could indeed speak, there would be no need for any science. We could just listen to what the world told us. On the contrary, science requires the patience to establish just what the facts actually are: something that is usually far more difficult than one might expect.

Scientific inference is the process whereby any description of the world and how it functions – usually referred to as a *theory* – must be tested in some way against empirical evidence. Inference takes place in our heads but uses empirical data from the world outside them. We accept or reject theories not on the basis of the authority, prestige or intelligence of their author, but on their ability to account for empirical evidence that we observe. By definition, we can never 'prove' a theory completely because, even if it were totally consistent with all the evidence anyone had ever discovered, we cannot know what evidence might arrive in the future. Because the evidence we have is incomplete, the conclusions we can draw are usually provisional or risky, in the sense that they may be wrong or that new evidence may persuade us to revise or improve them. Indeed, it makes sense to make such *falsification* one component of what counts as knowledge. Claims or statements that cannot be falsified count only as articles of faith. Scientific inference not only places a premium on the critique of existing knowledge but also requires the assimilation of new findings to some cumulative body of knowledge. What might at first appear to be a recipe for disaster – we know *nothing* with certainty – is actually a means to ensure that knowledge can accumulate and improve over time: it is open to growth. It does place stringent demands on the way in which evidence is produced and interpreted, but the pay-off from this effort is astounding. The development of knowledge, material progress and standards of living was extremely slow between the Neolithic revolution and the 17th century. The scientific revolution changed all that.

Statistical inference

In the social sciences, we use evidence that is systematically incomplete because it comprises data drawn from samples. Almost all of our knowledge of the world comes from them, because the social and natural worlds are just too vast, too complex and too dynamic to measure directly. Not only is it slow and expensive to collect information on millions of individuals, such a mammoth study would waste resources, and almost certainly generate data of poor quality. It would be very difficult to ensure that the measurements obtained were consistent and the data quality was good when collection requires a veritable army of enumerators. There would not only need to be checks, but checks on the checks, and checks on . . .That is why population censuses are usually only carried out once a decade and restricted to a very short list of questions. The last census in the UK cost only 87p per person per census year: a remarkably small amount considering the volume of information obtained, the logistical difficulties of attempting to track down everyone on census night and the marathon effort of checking the data collected and correcting the errors. However, it works out a total cost of £480,000,000: equal to the capital cost of a large hospital. The estimated cost of the 2020 US Census was $16 billion. In the natural world measuring populations, whether of animals, rocks, atoms or molecules, would simply be impossible. Some measurements or tests destroy or damage what is measured. Components may be quality tested to destruction, to assess their resilience. I'm happy for doctors to take a sample of my blood. I'd protest if they insisted on examining *all* of it. Finally, our interest is often in future populations, which cannot be measured for the simple reason that they do not yet exist. We have questions such as 'What would be the effect of . . .? or 'What would happen if . . .?' However, we may be able to generalise to such future populations from samples drawn from an existing one.

In all these situations, the sample data is of little interest in itself. Who cares what a small random sample of subjects might do or think? It is the population that is our real interest. But if we can generalise from our sample to the much larger population that it represents, we have a tremendously powerful analytical tool. Perhaps the greatest discovery of 19th-century science was the mathematical logic of how to infer the characteristics of target populations from random samples drawn from them. Because our data is limited, we need a good theory of how much we can *generalise* from the data we have observed in our research, to the wider world from which that data was taken, or to worlds which do not yet exist but lie either in the future or only ever in our imagination. The logic of how to do this is *statistical* inference. In its simplest form, it asks one of two questions:

1   Is what I have observed in my sample also what I *would* find in the target population if I could measure it?
2   If I assume the target population has feature X, how likely is it that I would see the data I observe in my sample?

Statistical inference is used to decide whether a pattern discovered in one batch of data that we have analysed is one that we would also be likely to find in other data that we have not, and often could not have, collected. The batch we examine is usually some kind of sample drawn from a target population. Often, we use these questions to do one of two things. We might want to produce an *estimate* of the size of something. What proportion of this population are women? What are the average earnings of social science graduates? At what age do women tend to have their first child? Or we might want to test whether or not some proposition about the population is true. We call such a proposition a *hypothesis*. Do science graduates earn more than social science graduates? Is average age at first birth older for women now or 20 years ago? Are those who identify as white more likely to vote Republican than others?

Inference can be used to travel in opposite logical directions. We can use the observed characteristics of individuals who comprise a random sample to infer information about the populations from which those individuals were drawn. We cannot observe or measure the whole population, but we can observe and measure the individuals. Conversely, we can use information we have about populations to infer otherwise unobservable or contested characteristics of individuals. For example, if we know the overall rate in a population for which a medical test returns a correct positive or negative result for an illness, and if we know the prevalence in that population of that illness, we can infer the probability that a specific *individual* has that illness, given their individual result for that test. Or in a court of law, if we have information about the general probability of some combination of factors coming together, or the prevalence of some state of affairs, we can make inferences about the probability of a very specific one: the guilt or innocence of a person charged with a crime.

Since inference is so important, and promises so much, it is neither surprising that the logic of scientific inference took some time to develop (it matured over about three centuries), nor that many people still find it challenging to understand today, nor that scientists continue to argue about the best procedures for drawing inferential conclusions. One reason for this is that it takes us out of our everyday world of intuitive semi-automatic inference production to one in which we compare carefully structured data with what we might *expect* to see if the world behaved in a certain hypothetical way. A clear head is needed to negotiate the logic of doing this.

This challenge is compounded by another one. As we shall see, the logic of *statistical* inference usually requires us to state precisely and explicitly any theory or hypothesis that we wish to test *before* designing the research and collecting the data. In most medical research, such preregistration is now an explicit requirement. Many people find this aspect of the logic of statistical inference the most difficult to grasp. Surely it must make sense to adapt research designs in the field, learning from experience. Surely it makes sense to learn from the data obtained, exploring and describing it, rather than restricting oneself to a hypothesis drawn up before the data was even observed! What difference can the timing of hypothesis generation possibly make? I can still remember my intense suspicion as a student that statistical logic must be badly awry when I was taught that data could only test hypotheses made *before* the data was observed. What harm could some gentle time travelling do? Why not generate hypotheses *post hoc*? Why should the timing of the generation of the hypothesis have anything to do with its validity? And what's wrong with multiple hypotheses? Surely social processes are sufficiently complex to demand more than one insight at a time about how they operate?

We shall examine this logic in Chapters 3 and 4, but the short answer is, unfortunately, all the difference in the world. This is perhaps the most fundamental contribution of the logic of probability to scientific inference. *Before* I witness the results of an experiment, or the data in a survey, or any other kind of measurement, I can use my hypothesis to calculate the probabilities of observing various results, if my hypothesis were true. It then follows that if the results, data or measurements I observe are very improbable given my hypothesis, then I have found good evidence to reject it. Conversely, if they have a high probability, I have some evidence that is consistent with my hypothesis. However, *after* I witness the results, data or measurements, there is nothing probable about them anymore. They simply exist. I can describe them in any way I like, but I cannot retrospectively calculate a probability that I can associate with any aspect of them. This does not mean that such data exploration is useless. On the contrary, such data exploration constitutes the bulk of empirical research in the social sciences. However, it does mean that the same batch of data cannot be used both to generate and to test hypotheses.

The need to start with theories and hypotheses implies that the next stage of inference is research design and the question 'What data do I need?' Often this question becomes 'What data can I get?' There is nothing wrong with bending one's research design to accommodate evidence that is available. However, there is everything wrong with thinking about these questions only *after* the data arrives. Naive researchers often have faith in what I like to think of as statistical 'magic dust'. It is sprinkled on data to find out 'what it says'. Sprinkle this dust on the data, and behold, it speaks! This approach to data is not confined to ingénue undergraduates. It is a fair description of the approach of many corporations to 'big data'. The moral of the story is this. The research question always comes first. Then come theories or arguments about what new evidence, given what evidence already exists, might offer promising routes to answering that question more effectively. Then comes a range of substantive research design questions around data collection and data quality. This is the stage, long before any data has been collected, that inference issues need to be sorted out. Among the most fundamental ones will be the following:

1 What is ethical, and consistent with informed consent, confidentiality and data protection legislation?
2 [For observational studies] what variables do I need to measure? Can the important prior or confounding variables be captured?
3 How is measurement error to be minimised or mitigated?
4 Is the data adequately random, or are there inevitable selection effects or other biases operating?
5 How much data is needed to give the study sufficient statistical power, keeping in mind the general trade-off between data volume and data quality, or the need to over-sample small groups in the target population?
6 Will the study be exploratory, or are there specific hypotheses to be tested? If so, have the procedures and data handling protocols been set out? Is pre-registration appropriate?
7 How will the data and results produced by the study be made available to other researchers for further analysis, meta-analysis, reproduction or replication?

While failing to appreciate the dangers of *post hoc* hypothesis generation is a danger we need to avoid, peril also lurks, unfortunately in concentrating too much on testing and failing to get familiar *enough* with the data. This means that data exploration is an essential companion activity to inference!

Exploration and inference: detectives and lawyers

The terms *inferential* and *descriptive* statistics are commonly used, often with the implication that they are distinct activities, and worse, that only the former is the serious stuff that merits attention. This is wrong. There are few, if any, research questions where careful description of the data does not play an essential part in drawing inferences. In many cases, such description may be more insightful and useful than a carelessly interpreted test statistic. There is no substitute in good research for getting thoroughly intimate with both the data and how it has been produced. No data is perfect but much data is 'good enough'. Knowing it well is the only way to judge just what it is good enough for. This approach to data analysis, first championed by John Tukey, is sometimes called 'exploratory data analysis' (EDA).

Much data is also 'noisy', in the sense of carrying a signal buried within many sources of error that tend to obscure it. As I emphasise at several points in this book, most measurement is a surprisingly difficult, uncertain and expensive process. Its results are inevitably imprecise. By contrast, the fixed, precise numbers in any data set create the opposite impression. Perhaps the most important skill in inference is the ability to keep in mind the origins of any data that is used and not be seduced by the mathematical precision of the numbers in a data matrix into thinking that it is more robust than it actually is.

In trial by jury, evidence is presented by defence and prosecution. Arguments are made about the interpretation of the evidence and whether it is consistent or not with innocence or guilt of the defendant. The latter is given five key advantages. First, the charges are specific. A defendant cannot be prosecuted for 'breaking the law'. They must be accused of something concrete, and they cannot be charged for one crime and then found guilty of something quite different. Second, the mere consistency of the evidence with the possibility of guilt is not enough, nor even the balance of probability. Proof must be 'beyond reasonable doubt'. The burden of proof rests with the prosecution, even if the word 'reasonable' begs a whole series of questions. Third, the proceedings of the courtroom are highly constrained in terms of what evidence is admissible, who gets to argue what and how the judge directs the jury about legitimate and illegitimate inferences and deductions that might be made. Fourth, no one with the remotest personal interest in the case can serve on the jury. Finally, defendants can appeal if they can show that the due process has not been properly observed. Despite all this, of course, dreadful miscarriages of justice do sometimes occur, especially when juries and judges share some strongly held convictions bearing on the case.

So it is with hypothesis testing. Good research design ought to take any personal interest of the scientist out of the equation. Transparency should ensure that procedures are sound and seen to be so. False positives usually ought to be more harshly dealt with than false negatives. The burden of proof should indeed rest with the scientist who makes a claim to new knowledge. Bogus knowledge is difficult to unpick; real findings can surface again if they are missed the first time. The evidence to securely establish new knowledge can sometimes take some time to accumulate. Both *post hoc* hypotheses and multiple comparisons are rather like trying to find the defendant guilty of *something* even if innocent of the original charge. This might be justifiable in some circumstances, but as the economist R.H. Coase argued, 'If you torture the data enough, nature will confess'.[1] Such confessions are unlikely to be of any value.

However, if the equivalent of trial by jury was all there was to statistical inference, it would play the same rather small, albeit important, part in research that jury trials do in the maintenance of the rule of law. The court proceedings may be the highly public and dramatic end of the legal process, but a far larger volume of investigative work precedes them: collecting and sifting the evidence, interviewing witnesses, identifying and locating suspects. Even the best court, fairest judges and intelligent jurors will deliver miscarriages of justice if the right evidence is not before them! Marsh, Tukey and many others have made the comparison between such detective work and 'exploratory' data analysis. As I show in Chapter 4, such work is an integral part of drawing reasoned conclusions from data, but only rarely does so in the form described by classical statistical tests. Indeed, such analysis has become more important as data collection moves to omnibus style surveys rather than bespoke enquiries.

The NHST wars

I used to run an email group for people teaching statistics to social scientists. It was fairly quiet: earnest veering on straightforwardly dull. But one subject was guaranteed to get list members pounding their keyboards in anger: null hypothesis significance testing (NHST), which is a core part of frequentist statistical inference. Many argue that NHST has caused a crisis in some scientific fields because the unthinking application of its procedures by scientists, journal editors and peer reviewers has led to the publication of unsound findings and the non-publication of other results that deserve attention but fail the NHST rules. A growing number

---

[1] According to Coase himself, in a talk at the University of Virginia in the early 1960s.

of statisticians are concerned that poor inference habits have become so ingrained as to undermine the scientific character of research and led to systematic biases in the nature of the organisation and reporting and publication of scientific work. Eventually, in 2016 and again in 2019, the American Statistical Association (ASA) along with the world's leading science journal (*Nature*) questioned the way significance levels and *p*-values were being used and in 2019 argued that the term *significance* itself should be abandoned. Feelings have run high. Whoever would have thought that inference was so exciting?

There is some, if not much, justification for this. Statistical inference inevitably requires making 'risky' statements that, because they are based on finite empirical evidence, are only probably correct rather than true or false by logic or definition. There is no absolute 'one best way' to do this, and given the nature of probability, even the 'best' method must sometimes fail. Statistics, as David Spiegelhalter (2019) has recently reminded us, is an *art*. However, what is at stake is the integrity of science. Most science is underpinned by drawing inferential conclusions from data. The validity of these conclusions is tested or proved through scientific debate and replication (or failure to replicate) to weed out rogue results, leaving the secure ones standing.

However, scientists arguing about inference present a problem for students. Inference requires some concentration, effort and practice to learn. It is discouraging to discover that once mastered, NHST delivers results that turn out to come with a less than cast-iron guarantee! Partly because of the way higher education is run, students usually want to know what is the *correct* result or right answer. Chapter 6 discusses the NHST wars, but it may help the reader to anticipate some of that discussion in order to explain why a thorough understanding of the logic of NHST is indispensable, regardless of what view one might eventually take of NHST itself.

As Chapter 3 explains, inference often takes the form of formulating a null hypothesis about a target population and using sample data to calculate the probability that we would observe what we see in the sample, conditional upon the hypothesis being true in the target population from which the sample has been drawn. Typically, the null hypothesis is some statement such as there is 'no difference' between two different parts of the population on some measure, or that the value of some parameter is zero, or that some treatment has no effect. Often the aim of inference is to see if we can *reject* the null hypothesis so that we have evidence that some difference does exist or that a treatment does have some effect. We formulate a null hypothesis not because of some perverse delight in double negatives, but because it is often the only quantity that we *can* calculate and estimate. As we have just noted, null hypotheses have to be formulated *before* the data is observed, because once we have sight of the data, there is no longer anything probable about any pattern we observe in it, or any hypothesis consistent with that pattern that we might want to formulate.

This creates four problems. The first is that of multiple hypotheses. Just how many hypotheses can we test with some data? Unfortunately, random variation means that the more questions we ask, the greater is the risk that some chance pattern in the data flags up a positive result that would not be replicated in the population. This problem is compounded by the industrial scale of modern science. I might ask the data only one question. But hundreds, or even thousands, of my colleagues might pose others! The second is that of statistical power. Given enough data, there is virtually no null hypothesis that cannot be disproved. However, the effect or difference measured may have no practical or scientific value. Conversely, with insufficient data it is easy to fail to reject a null hypothesis and keep it intact, when in fact we would dismiss it if we had more adequate evidence. The third problem is the kind of question we can answer. As we shall see, at the core of NHST is the calculation of the probability of observing the sample data, conditional upon the null hypothesis being true in the population. This probability is often a very useful piece of evidence, but it is not what we would often ideally like to know, which is the probability of the null hypothesis being true in the population, conditional upon observing the data we obtain. This latter probability can usually only be calculated if we already have some estimate (a *prior* probability) for the truth of the hypothesis. In some situations, it would be stupid not to take such priors into account, but in others, it violates the basic scientific credo of 'Nothing on another's word': the motto of the Royal Society that encapsulates the appeal to empirical evidence as superior to any other authority. The fourth problem concerns what we do with our hypothesis test result. As we've already noted, the latter is a value between 0 and 1 for the probability of observing our data if the hypothesis is true, but we often want to reach a verdict or take a decision based on this number, which produces the problem of where any dividing line ought to be drawn.

With a good command of the logic of inference, it is not difficult to deal with these four problems. Here we come to the nub of the issue however. Understanding inference imposes a 'cognitive load'. It is difficult. The mathematical logic underpinning it is not too complex, but it easily defeats the first efforts at understanding of anyone not used to thinking in logically rigorous terms. It requires a clear head and some concentration. Taking even a few steps along a chain of logic is difficult the first few times you do it. It takes practice to understand each step well enough to be able to keep some attention on the journey as a whole. This is compounded by the messy nature of empirical reality, which is often challenging to reduce to the probability models that are the bedrock of all inference. It is often hard to see the link between rolling dice or flipping coins and getting good evidence of how social inequality is structured or what model might offer the best description of voting behaviour. Because of this, courts of law have sometimes witnessed grave miscarriages of justice based on the misunderstanding of inference.

Scientists usually have more interest and motivation to tackle the substantive questions of their chosen discipline than to devote their energy to statistics. Because it takes some time, concentration and practice to gain a thorough understanding of the basic principles of inference, the tendency has sometimes been to fall back on ready-made NHST-based 'cookbook' recipes for inference that can be rapidly rote learned and applied to almost any research problem. Software now takes care of any calculation necessary and offers various inferential tests at the click of a mouse. The trouble is that in drawing inferences, evidence comes in many different shapes and sizes, as do the questions that the evidence has been produced to address. In this situation, 'one size fits all' may give a formally correct answer to the wrong question and can lead researchers to draw quite mistaken conclusions from their evidence.

Now assume that these statistical shortcomings are shared by the working scientists and by those who evaluate their work: journal editors, peer reviewers and research funders. The result could be a chronic problem of poor science. This has been compounded by the way in which scientists tend to work within disciplinary silos. Human beings are generally not very good at planning the future in detail for the simple reason that it is very difficult to do so. Yet good research requires just this ability: to work backwards from precisely what data is needed to answer some question to the entire process of producing that data. What will happen within each stage of a research project is subject to all manner of eventualities, some reasonably foreseeable while others not. Each resulting open possibility multiplies the potential planning required exponentially. The easiest path through this methodological maze is often to do what colleagues have successfully done before. As a result, habits build up within disciplines. These habits may no longer have much rationale, but persist because they come to be what everyone is expected to do.

Pressures of career insecurity, the pursuit of status or prestige, or the simple average distribution of malice or mendacity in the scientific population can lead to corners being cut or selective interpretations of the data made. Journal editors or reviewers are human and have their own sets of prejudices, also known as sincerely held and genuine convictions, borne of long experience. Add to this mix the rapid expansion and industrialisation of science; the increasing commercialisation of higher education, including research; and the temptation to organise it with targets and all the paraphernalia of 'management'. A small community of scientists can be bound by a tight and demanding moral code of professional integrity. Not so a whole industry.

The solution to these challenges lies not in any new theory or changed procedures of statistical inference but in a sounder grasp of its fundamentals, which this book aims to provide in a language accessible to those with no formal training in statistics.

Inference, reproducibility and replication

Scientists sometimes approach inference as a means to squeeze the maximum possible amount of information from the minimum amount of data. There are situations when this is the best we can do, but they are rather unusual. Rather than torturing the data till it offers up some secret, it is often better to collect some more. James Steiger (1990) put this well when he commented, 'An ounce of replication is worth a ton of inferential statistics' (p. 176). One of the strongest tests of whether some feature of a sample is also likely to exist in a population is to see whether it can be found in *another* sample. The internet facilitates the curation and distribution of data, making it more accessible than ever before. A simple, but vital, test of any analysis is to ensure that others can repeat it and obtain the same results. The constraints and costs of hard copy publication used to limit the volume of research that print journals could report. But the internet has abolished that constraint. There are now few technical barriers to making the entire process of any data analysis thoroughly public, rather than only the main results. Replication, meta-analyses and other scientific collaboration will both become easier and more necessary in the social sciences in the future. To the extent that science becomes more collaborative and collective, it is also likely to become more transparent and less open to error or abuse.

Inference in action: the Salk Vaccine trial

A good way to appreciate inference is to see what it can do. An excellent example of its application is the Salk Vaccine trial, summarised by Freedman et al. (2007, pp. 3–6) and Dawson (2004). Urbanisation and improved water supplies and hygiene in affluent countries caused the incidence of polio to increase in the first half of the 20th century. Polio mostly struck children. Many cases might result in little more than a fever, but a small proportion went on to cause paralysis, life-changing disabilities or death. The UK reported between 2000 and 10,000 cases of polio each year, while in the USA, in 1949, there were more than 40,000 cases. Some opinion polls, early in the Cold War, ranked the fear of polio as second only to the fear of nuclear annihilation. While it was known that a virus was responsible for the disease, its origin and means of transmission were unknown, and the virulence, location and timing of epidemics were unpredictable. Cases seemed to concentrate in the summer months, so that a link to ice cream was suspected. Another suspect was the sitting posture adopted by children at school!

A vaccine was developed, but it was vital to know whether it worked. However, nobody could know in advance which children might be struck by polio or where they might live. Thus, in order to test the vaccine, a vast study was organised in the USA in 1954, in which almost half a million children took part. Because so little was understood about the nature or mechanics of polio transmission, the study had to be nationwide. As far as possible, it was important to have all parents let their children participate in the study in order to avoid bias by class, region or other factor. Careful blinding was used so that no one but the researchers knew which children had been vaccinated and which not. Children were randomly allocated to the vaccine and placebo. In the course of the trial, 142 children who received the placebo and 57 children who received the vaccine went on to contract polio.

Did the vaccine work? Here our informal inferential skills tell us almost nothing. We are clearly dealing with probabilities rather than certainties. Fifty-seven vaccinated children nevertheless contracted polio, while the vast majority (>99.9%) of *un*vaccinated children did *not* contract the disease. I doubt that many courts of law asked to rule on whether the vaccine was 'guilty' of preventing polio would have decided that this had been proved 'beyond reasonable doubt'. Our intuitive plausible model might be that perhaps the vaccine made some difference, but the ratio of between 2:1 and 3:1 between the incidence of polio in the vaccinated and unvaccinated children might lead us to conclude that the difference was unlikely to have been much. Perhaps it might be best to concentrate research efforts elsewhere, for example, on that possible link to ice cream.

However, one method of statistical inference (the chi-square test, which is described in Chapter 4, Section 'Chi-Square and Contingency Tables') allows us to calculate the probability of getting the result in the Salk trial if the vaccine was indeed useless. It hinges on the fact that randomisation was used to decide which of the children studied were vaccinated and that neither the children receiving the vaccine, nor their parents nor the doctors responsible for diagnosing cases of polio knew whether they had been vaccinated or not. These precautions ensured that the *only* difference, on average, between the two groups of children was the fact of vaccination. *Because* of this appropriate research design, the function of the vaccine could become a straightforward statistical question: what was the probability of getting 142 polio cases in one group and 57 in the other simply by chance, which would be the case if the vaccine had no effect. The chances of getting this size of difference between the rates of polio in the two groups by chance were very small: about one in a billion. Our intuition alone could never have worked out that result!

However, note two things. Scientific inference worked hand in hand with statistical inference. Without randomisation, double-blind vaccination and polio detection procedures, the data could not have been interpreted in the same way. I've also described the results in a rather legalistic way. I did not write 'the study proved the vaccine worked'. The vaccinated children who contracted polio would have had reason to complain if I did. The chances I described compared the data with results

I could have expected to obtain had the vaccine had no effect at all. I didn't actually say anything about the vaccine itself or about what performance we'd expect a vaccine to reach before we described it as 'working'. The latter might bring in all kinds of other considerations apart from the estimate of its rates of prevention. Does it have side effects or other risks? Is it costly or difficult to administer? However, the Salk study clearly showed that vaccination was potentially effective and a promising line of research. More effective vaccines were rapidly developed, so that within a few years, rates of polio plummeted. In 2018, just 33 children on the planet contracted polio, which had been eradicated in every country in the world except Nigeria, Pakistan and Afghanistan. Statistical inference was only one part, but nevertheless a vital part, of such progress.

Inference in action: fertility and development

In the course of the 1950s and 1960s, policymakers in the USA became concerned about world population growth. One fear was that if such growth outran economic growth in the developing world, the resulting poverty would fuel support for radical and communist parties and increase the influence of America's Cold War adversary (Connelly, 2008). Another fear was that of a global Malthusian population crisis with widespread famine driving civil disorder and warfare (e.g. Ehrlich, 1968). These fears led to the USA making action on family planning a condition of development aid. In turn this led some governments, including that of India, to promote sterilisation and other contraceptive methods with little concern for human rights, leading to widespread abuse.

Fear of 'overpopulation' took far too little account of either innovation in agriculture or the impact of economic development on fertility. It soon became clear that 'development is the best contraceptive' as developing countries had argued at the 1974 World Population Conference in Bucharest. Analyses of fertility surveys were able to show that factors such as urbanisation, industrialisation, access to education and greater political rights for women were associated with dramatic and rapid fertility decline. The fertility rate for the world as a whole is now estimated to be 2.4 children per woman, half of what it was only 50 years ago and around 30% lower than the rate for the USA at the time fears of overpopulation reached their height. No one has ever asked every adult woman in the world about where they live; whether they, or their partner if they have one, are employed; their level of education or years in school; how many children they have had or their contraceptive behaviour. Surveying more than 7 billion people would be a gargantuan task and a largely useless undertaking. However, information taken from samples allows us to make confident estimates about much larger groups of women, again using randomisation.

Finally, the inference that problematic over-population was an unlikely scenario required neither hypothesis testing nor a thorough understanding of the determinants of fertility. In fact, there is extensive and inconclusive debate amongst demographers about the latter. Inference simply required some careful measurement in order to estimate and describe the probable general trend of fertility.

The world before statistics

Remarkably small samples of data can be enough to give us the information we need about very large target populations, but this insight is barely a century old. Without such statistical inference, our knowledge of the world would be much poorer, and our ability to act effectively within it much less. Not only would it be much harder to develop effective medicines, we'd struggle to understand or control economic conditions, we'd know almost nothing about our societies or the social institutions and processes they comprise.

Such a world only began to disappear a little over two centuries ago. Over the course of the 19th century 'state-istics' slowly developed, alongside the realisation that quantitative information about social, economic and political conditions was needed for any kind of domestic management. Population censuses became a tool used by all states, but they were unwieldy and expensive. Over the course of the 20th century, governments and individual research teams discovered the virtue of the sample survey and gradually refined its organisation and methods; however, beyond some key economic data, there are remarkably few comprehensive good quality surveys of social and political conditions carried out before the 1970s.

The world before statistics was also an overwhelmingly static one. Although individuals might have tremendous practical knowledge about some details of their immediate natural environment, ignorance and prejudice prevailed when it came to turning that knowledge into any kind of system so that any innovation or change that occurred was piecemeal and marginal. An ordinary person living at the time of the Roman Empire, transported by time machine to the 15th or 16th centuries, would not have found much to amaze or disturb them. As is well-known, science, industrialisation and seemingly inexorable economic progress have transformed the world since then. A time machine trip of a few decades would be enough to disorientate and amaze most people. My daughter teases me by pretending disbelief that I could be so old as to remember a world without either mobile phones or the internet.

What is sometimes less well appreciated is that one of the essential preconditions of this colossal transformation was the invention, development and diffusion of scientific thinking and logic. This diffusion has become so thorough that we find it difficult to imagine a world in which they are not fundamental, even though their origin is so recent. However, this diffusion does not mean that all parts of the core of scientific logic are well understood, including by scientists themselves. Anyone wishing to understand the modern world therefore needs to understand scientific logic. Statistical inference plays a vital part in that logic, because only through inference can we use small and tractable samples of the world to infer conclusions about and understand the world as a whole. Anyone wishing to understand the modern world therefore needs to understand inference. That is what this book is for.

What knowledge I assume

Any methods book can either start from scratch, risking the boredom of readers who already have some knowledge, or make some assumptions about what readers already know that puts it beyond the reach of others. In the rest of this book, I assume that readers have some general familiarity with quantitative methods; understand the language of variables, values and cases; can follow its presentation in frequency or contingency tables or graphical representation by bar charts or histograms; and have enough maths to cope with fractions, decimals percentages, powers and square roots. Any introductory data analysis or statistics text covers these issues.

The structure of this book

Chapter 2 introduces the formal understanding of probability, which underpins all inference. Chapter 3 sets out the logic at the heart of inference from first principles by the thorough analysis of a simple experiment. I am sure that some readers will wonder whether and why it is necessary to go into this experiment in the length and depth that I do, and will be impatient to get on to the more 'applied' chapters which follow. Please resist this urge. The principles that I set out through using this example are the foundation of all reliable inference, regardless of the subject matter, whether the approach used is 'frequentist' or 'Bayesian', whether the data used is 'big' or otherwise, or produced by experiment or observation. It introduces every important concept in inference. Digesting and understanding this chapter makes the rest of the book far easier to follow. However, much more important than this, by providing a clear exposition of the logic of inference, it equips you to follow and participate in the lively debate about just how inference ought to be undertaken in scientific enquiry, as well as the relative importance of data exploration and analysis, statistics, research design, reproducibility and replication. Chapter 4 extends the analysis presented in Chapter 3 to the wide variety of statistical tests that readers are most likely to encounter in the social scientific literature or need to employ themselves, as well as discussing when *any testing at all* is appropriate. Chapter 5 deals with regression specifically, because it occupies such a large part of social science statistical analysis. Chapter 6 deals with power, effect sizes and 'inverse probability', and Bayes rule. Finally, Chapter 7 reviews the NHST debate in a spirit of clarifying issues rather than taking sides, but making it clear that NHST will continue to play an important role in statistical inference, and placing inference in the context of the changing organisation of science.

Chapter Summary

- Any use of empirical evidence to reach some conclusion requires making inferences, which are almost always uncertain.
- Statistical inference estimates how likely it is that empirical evidence based on a sample drawn from a target population enables us to draw conclusions about that population.
- Our intuitive perception of the world around us tends to comprise unreliable causal narratives that we may unconsciously use to confirm our existing beliefs.
- Scientific knowledge usually takes the form of using existing knowledge to formulate and test the consistency of hypotheses with empirical evidence.
- This creates a tension between exploring data to generate new hypotheses and using data to test those hypotheses.
- Poor understanding of the logic of inference, or the rote application of rules used in null hypothesis significance testing has compromised some areas of scientific knowledge.
- The best solution to this problem is a secure knowledge of the underlying logic of statistical inference.

Further Reading

Hacking, I. (2001). *An introduction to probability and inductive logic.* Cambridge University Press.

This book is an excellent introduction to the general logical and philosophical issues, setting the statistics in the general context of making 'risky statements' and written in an engaging and lively style.

A more general introduction to inference and statistics is David Spiegelhalter's (2019) *The Art of Statistics.* An introduction inspired by Tukey's approach is Cathie Marsh and Jane Elliott's (2008) *Exploring Data*, 2nd edition, Polity Press.

2

PROBABILITY, RANDOMNESS, PROBABILITY DISTRIBUTIONS AND SAMPLING DISTRIBUTIONS

Chapter Overview

What is probability?

You are already entirely familiar with probability. When you have thoughts like 'it looks as if it will be windy today', 'there must be 10,000 people watching this volleyball match!' or 'there must be intelligent life elsewhere in the universe', or when you decide to walk because it will be quicker than waiting for the bus or select your favourite coffee brand because it tastes better, then you are dealing with probability. Even if the sky is dark and foreboding, the clouds may pass, and the storm may never arrive. The number of people watching the volleyball match is definite, but both how good your estimate of that number is and how confident you are about it are matters of probability. Nobody knows if there is intelligent life elsewhere in the universe, but we may each have our own subjective idea, ranging from extremely unlikely (the preconditions for the existence of life are extremely exacting) to almost certain (if there are countless trillions of planets, at least one surely supports some form of life). Such life either exists or does not: probability can only express our degree of belief. You don't know if you have just missed the previous bus, or whether the next arrival is imminent, so you make a guess of how long you *probably* have to wait and whether it is worth it. Again, the bus will eventually arrive at a definite time, but until it does, we cannot know for certain exactly when that will be. You definitely like your favourite coffee, but who knows, there might be an even tastier one out there.

Probability thus deals with descriptions or predictions that may or may not be correct, or outcomes to processes that are not strictly determined, but instead have some random component. That random component may only represent some incompleteness in our knowledge. Thus, when I deal some playing cards from a well-shuffled pack, the randomness is a function of my ignorance. Were I able to see each card in the pack as it was shuffled, I would know which cards were going to be dealt. Conversely, it may represent the impossibility of gaining such knowledge. Were I watching someone spin a roulette wheel, or roll some fair dice, even the most extensive analysis and knowledge of things, like the speed at which the wheel was spun, the direction of the dice and so on, would leave me facing the inherent randomness of the process. Randomness is essential.

It is also therefore important to distinguish probabilistic insight from hindsight. *After* I have drawn a card from a pack, flipped a coin or rolled some dice; or *after* a day has taken place; or *after* the bus has arrived, there is no longer anything probable about it. The card I obtained, the faces on which the dice or coin landed, or whether the wind blew or when the bus came are all fixed. The physicist and Nobel laureate Richard Feynman used to make this point by announcing the following at the start of a talk:

The most amazing thing happened to me tonight. When I parked my car, the registration number of the vehicle in front was [whatever number or letters you please]. What was the chance of *that!*

The joke, of course, was that chance had absolutely nothing to do with it! Had Feynman predicted the registration plate *before* parking that *would* have been impressive. The uselessness of 20/20 hindsight might seem an obvious point. However, in the heat of data analysis and inference, it can often get missed, with dire consequences, as I discuss in Chapter 4.

Thinking about the world in terms of probabilities, instead of causal narratives, takes some effort and practice. Our brains do not seem to be well evolved to deal with randomness and probability. You can confirm this with a simple thought experiment. Before reading on, choose a number at random between 1 and 10. Once you have done this, first reflect on how difficult it was. Did you, for example, discount 1 and 10 from your choice because, being at the start and end of the sequence of numbers, they did not feel very random? Perhaps you discarded 5 for the same reason, as a number in the middle. Every time I've done this exercise with a group of people, about one quarter or more choose the number 7. If people were capable of random selection, this number, like every other one, ought to come up only about 1/10th of the time. Even in a small group of 50 people, the probability of 12 of them choosing 7 randomly is vanishingly small: less than 1 in 500. People plump for 7 as it has so many other connotations (days of the week, Wonders of the World, Snow White's dwarves etc.). Our brains simply cannot 'do' randomness. So how do we measure probability?

Human beings have played games of chance for millennia. Early dice were formed by using knuckle or hoof bones of animals that were approximately square. However, until barely three centuries ago, the idea that the outcome of unpredictable events might be approached through any kind of numerical analysis was never developed. Why it emerged when it did is something of a mystery, as several plausible explanations can be discounted. Until it did so, people were apt to divide the world into that which was known and predictable (the ordered world of measurements and natural 'laws') and that which could not be known (or known only to the gods): fate, or what Machiavelli called 'fortuna'.

At first sight, this appears to be a perfectly sensible proposition. There are things and processes that we can describe and predict because we have measured them. The study of physics, chemistry and biology is full of natural 'laws' that describe not how matter 'might' behave but how it always does behave, so far as we are aware. If I drop a stone, it does not have a 'high probability' of falling to the ground under the influence of gravity: it *must* fall. So far as we can tell, light travels at the same speed in a vacuum everywhere in the known universe: there is nothing probable about its speed. On the other hand, there are things and processes that no person, no matter how well informed, can predict or describe perfectly. Machiavelli was only too aware that no matter how well prepared a prince might be, 'fortuna' might finish him off. Harold Macmillan (UK Prime Minister, 1957–1963) was reputed to have described the main threat to a government as 'Events, my dear boy, events.' No record has been unearthed of him ever saying this, but the statement rings so true that it has acquired the status of an urban myth.

'Events' are the basic building blocks of a scientific approach to probability. Our understanding rests on a simple paradox. Individual random events are, by definition, unpredictable. This makes randomness frustratingly difficult to produce artificially and explains the often bizarre mechanical and electronic contraptions fashioned by those who organise lotteries to produce genuinely random numbers. If I choose a lottery ticket, spin a roulette wheel, draw a card from a properly shuffled pack, flip a coin or roll some dice, there ought to be no possible way to predict the single event that is the outcome. If I discovered one, I would be very rich indeed. Conversely, if I flip a fair coin a couple of thousand times, I can be very certain indeed of the approximate share of *heads*. It will be very close indeed to one half. So too with the cards or roulette wheels or dice. Each face will land upwards a roughly equal proportion of times, so too will each number on the wheel, or a card in the pack. This is a good approximate description of 'the law of large numbers'. This states that the distribution of possible outcomes from a large number of individually unpredictable events settles down towards a set of proportions corresponding to the underlying probability of each outcome. Individual events may be entirely unpredictable, while large collections of them can be entirely predictable. This is the essential paradox of probability. So how do we reckon the probability of events? First, we need to review some basic concepts.

Some elementary descriptive statistics and notations

If you are already familiar with basic statistics, feel free to skip this section.

Quantitative data always takes the form of the *values* that *cases* take for a *variable*. Cases can be people, organisations, regions, countries or anything that might be counted. Variables are any characteristic of cases that can be classified or measured. Data is usually stored in the form of a data matrix where each row represents a case, each column represents a variable and each cell contains the value for the corresponding case for the corresponding variable, as in the example in Figure 2.1. It shows the first seven cases from a data set with variables about the sex, nation of birth and weight at birth for a sample of children born in 1958. Thus, for example, baby ID number 004 was a girl born in Scotland weighing 3005 grams at birth.

ID	Sex	Nation	Birthweight
001	Male	Scotland	3657
002	Male	Scotland	1729
003	Female	Scotland	3487
004	Female	Scotland	3005
005	Female	England	3063
006	Female	Scotland	3175
007	Male	Wales	3969

Figure 2.1 A data matrix

Note. National Child Development Study teaching data set.

The first thing to note about variables is that they can vary in different ways. Birthweight can take a range of values that are infinitely divisible between the lightest and the heaviest weights possible. No matter how precisely we tried to measure any weight, we could always, in theory, make a still more precise measurement. Rather than do this, we typically record such amounts as intervals within which the true measurement lies. For example, a measurement of 100 grams 'to the nearest gram' implies a weight greater than 99.5 but less than 100.5 grams. Hence, the name given to this level of measurement: *interval*. Many interval-level variables, such as weight, also have a meaningful zero. When this is the case, they are sometimes referred to as *ratio* variables, since the ratio between any two values will have a clear meaning. If I say that one baby is twice as heavy as another, my meaning is unambiguous. If you do not have a real zero, such ratios become meaningless. Temperature measured in degrees Celsius does not use a real zero. It would make no sense for me to say that today is twice as hot as yesterday because today it is 30 degrees while yesterday it was 15 degrees. Measured in Fahrenheit or Kelvin, these numbers would be different, and the ratio would change.

Nation of birth is a *categorical* variable. The only form that variation takes is that we can put cases in different categories. We cannot make any meaningful comparison between these categories, apart from the fact of their difference. It would be nonsense to say that England was 'nationer' than Scotland, or that Wales was twice the nation of Scotland. Since all we can do is give the categories names, the term for this level of measurement is *nominal*.

You will sometimes come across a third level of measurement intermediate between these. *Ordinal* variables are ones where the categories can be ranked or put in *order*, but we cannot describe the relationship between these values in terms of a number, as we can for interval-level variables. Ordinal variables are often produced by questionnaires asking people about their (dis)agreement with something. Respondents may be asked whether they

- strongly agree,
- agree,
- neither agree nor disagree,

- disagree or
- strongly disagree.

These categories are not only different from each other, but we can also put them in a sensible order of progressively greater agreement. However, it would make no sense to say that 'strongly agree' represented twice as much agreement as 'agree'. We know it is more, but not *how much* more.

In the form shown in Figure 2.1, data is indigestible, so we summarise it in various ways. Two important summaries are the *level* (average) and *spread* (dispersion) of the values of a variable. Two useful measures of level are the *arithmetic mean* and the *median*. The *arithmetic mean* is calculated by summing the values for a variable across all its cases and dividing by the number of cases. For example, if these seven cases constituted our entire sample, the mean birthweight in grams would be:

$$\frac{3657 + 1729 + 3487 + 3005 + 3063 + 3175 + 3969}{7} = \frac{22,085}{7} = 3155$$

Note that if you subtracted this mean from each of the original variable values, you would obtain a set of *residuals*, each of which measured the absolute distance between the value for that case and the value of the mean, and that the sum of all these residuals would be zero.

$$502 - 1426 + 332 - 150 - 92 + 20 + 814 = 0$$

Often, prose descriptions of operations with numbers are cumbersome, or imprecise. It is therefore sometimes helpful to use special notation to keep things concise and precise. You almost certainly use such notation already, including:

+	add
−	subtract
%	divided by 100 or out of every 100 or per cent

In notation for statistics, we usually use letters such as X or Y to stand for variables and N for the number of cases. When referring to a population, we usually use upper-case letters. We use lower-case letters for a sample. Four useful symbols are:

Σ	'add up everything defined to the right of this symbol'. The symbol is the upper case of the Greek letter sigma
x^2	'multiply x by itself' = $x * x$, e.g. $3^2 = 3 * 3 = 9$
$\sqrt{x}$	'take the square root of x' = the number which, multiplied by itself, produces x, e.g. $\sqrt{9} = 3$; $\sqrt{4} = 2$
$\bar{x}$	the arithmetic mean of x, pronounced 'x-bar'

Using this notation, the formula that we have just seen for the arithmetic mean becomes:

$$\frac{\Sigma x}{n} = \bar{x}$$

And the formula for the residuals from the mean becomes:

$$\Sigma(x - \bar{x}) = 0$$

The *median* is calculated by ranking all the values for all the cases of a variable and taking the value of the middle-ranked case(s). For our example with seven cases, the value of the median is:

1729 3005 3063 **3175** 3487 3657 3969

Spread measures how closely bunched or spread out the values are. Corresponding to the arithmetic mean, the *standard deviation* is calculated in four steps. It is important to understand how a standard deviation is calculated, because the method, which involves taking *a sum of squares*, is one that underpins many other procedures that we will encounter later in the book.

1   The mean is subtracted from each value to produce a residual (Figure 2.2).

Value		Mean		Residual
1729	–	3155	=	−1426
3005	–	3155	=	−150
3063	–	3155	=	−92
3175	–	3155	=	+20
3487	–	3155	=	+332
3657	–	3155	=	+502
3969	–	3155	=	+814

Figure 2.2   The residuals

2   Each residual is squared. This has the effect of removing the sign and making every residual positive (Figure 2.3). They are added together to produce the *sum of the squared residuals* (often referred to as the *sum of squares*).

-1426^2	=	203,3476
-150^2	=	22,500
-92^2	=	8464
20^2	=	400
332^2	=	11,0224
502^2	=	25,2004
814^2	=	66,2596
Total		308,9664

Figure 2.3   The sum of squares

3   The *sum of the squared residuals* is calculated, and divided, by the number of cases:

(2033476 + 22500 + 8464 + 400 + 110224 + 252004 + 662596) = 3089664 ÷ 7 = 441380.6

This value is called the *variance*. Because we squared the residuals, the variance is in squared units.

4   To return to our original units, we take the square root of the variance to obtain the *standard deviation*.

$$\sqrt{441380.6} = 664.4 \text{ grams}$$

In statistical notation, the Greek letter sigma σ is used to represent a population standard deviation, so that we could write the formula:

$$\sigma = \sqrt{\frac{\Sigma(x-\bar{x})^2}{n}}$$

The standard deviation has many useful mathematical properties, and the procedure of squaring residuals is one that has many uses that we will see later on. It is therefore a good idea to become comfortable with the idea. You can think of the standard deviation as the 'average' distance of values from the mean value. Because of the squaring process, *outlier* values far from the mean have a disproportionate impact on its value. This is shown by the first case in our example above, where more than half of the variance comes from just one case (ID 002) with a weight a little over 2 standard deviations below the mean. If we are taking the standard deviation of a sample rather than a population, we correct for the fact that there is less variance in any sample than in the population from which it was taken by dividing by one less than the number of observations we have, so that the formula becomes:

$$\text{s.d.} = \sqrt{\frac{\Sigma(x-\bar{x})^2}{(n-1)}}$$

An alternative measure of dispersion to the standard deviation is the *interquartile range* (IQR). *Quartiles* take the value of the case ranked half way between the lowest value and the median, and between the median and the highest value. For both medians and quartiles, when we deal with an even number of cases, the value of the quartile is the average of the values of the two cases, which straddle the middle of the ranking. Thus, the lower quartile for our seven babies' weights is the average of 3005 and 3063 = 3034.

1729

3005⌉
3063⌋

3175

3487

3657

3969

The upper quartile will be the average of 3487 and 3657 = 3572.

1729

3005

3063

3175

3487⌉
3657⌋

3969

The IQR will therefore be 3572 – 3034 = 538 grams.

With these concepts under our belt, we can examine the mechanics of probability.

Trials and sample spaces

Instead of *events*, the analysis of probability uses the terms *trial, outcome* and *sample space*. A *trial* is any set-up in which more than one outcome is possible. We describe this more formally by saying that a trial has two or more *random* outcomes. Random outcomes can be events (a coin lands on *heads*, an election takes place, a person becomes unemployed), descriptions (female, aged 42, African) or any other trial result. We can list *all* the possible outcomes of a trial. This is called the *sample space*.

The sample space of a dice roll is [1, 2, 3, 4, 5, 6]. The sample space of a coin flip is [Heads, Tails]. The sample space of whether it will rain tomorrow where I live is [Some rain, No rain all day]. Outcomes in a sample space can be *discrete* categories like [Yes, No] or [Heads, Tails] or *continuous* such as measurements of length, income, spatial location, weight, time or any other quantity. Often we are interested in one particular outcome to a trial, referred to as the *outcome of interest* or *success*.

Any process or situation that is not fixed, constant or determined in advance can be thought of as a trial. This includes any measurements or descriptions we might ever make. Therefore, experiments, censuses and surveys are all examples of collections of *trials with random outcomes*. In a social survey, we could think of each question asked as a trial, and the potential answers to that question as the sample space of its random outcomes. Each time each question was addressed to a respondent, we would have a repetition of the trial. There are two rules that the random outcomes in the sample space of a trial *must* always follow:

- They must be *comprehensive*.
- They must be mutually exclusive.

This means that:

- the sample space must cover *every* eventuality that could arise,
- each time the trial takes place, it must end in *some* outcome and
- the same trial cannot have more than one outcome at a time.

In other words, *some*thing must happen, but two things cannot happen at once. This is ultimately common sense. A flipped coin must land on something (even if it is on its edge – estimated at 1 in 6000 tosses for a US nickel (Murray & Teare, 1993, p. 2547) and cannot land on both sides at once. If I ask about a survey respondent's age in years last birthday, they must have such an age (even if they do not know what it is), and the answer might be any number between 0 and a little over 100, but it could not be *both* 7 *and* 42 for the same respondent. The sample space for a survey respondents' reaction to the statement 'Gays and lesbians should be free to live life as they wish' might be

> [strongly agree, agree, neither agree nor disagree, disagree, strongly disagree, don't know, refuse to answer]

The sample space for a question about their income or height would be a continuous range of values in metres or dollars, or any other measures of height or money.

At first sight, these questions do not look like trials. There is nothing probable about anyone's age, their sex at birth, height, income or views about homosexuality. All these characteristics exist, just like Richard Feynman's number plate. However, all these definite characteristics are a function of countless random processes, including as a final stage their selection for measurement in some census or survey. What is probable about them is that *until the trial takes place*, we do not know what the outcome might be. It has the same status as the number plate on the car that might, or might not, be the one that Professor Feynman parks next to.

The law of large numbers

In order to calibrate the probability of any individual outcome from a trial, we use a scale from 0 to 1, where 0 means completely impossible and 1 means absolutely certain.

$$0 \leq P \leq 1$$

Frequentist approaches to probability are rooted in the *law of large numbers*, which states that if exactly the same trial is repeated many, many times, and the outcomes of the trials are independent of each other, then the proportion of each outcome approaches a limit that corresponds to the underlying probability of that outcome. The probability is the number or *frequency* of each outcome, divided by the number or frequency of trials. 'Independence' just means that the outcome of one trial has no effect on the outcome of other trials.

Most introductions to statistics turn to flipping coins, rolling dice, drawing balls from urns or playing cards at some point. This is not because games of chance are important, but because they provide simple and familiar models of random processes. If these games did not have a strong element of randomness, they would be less fun. The element of pure chance evens up the prospects of experts and beginners in the very short run, giving everyone the possibility of success and creating the phrase 'beginners' luck'. Imagine the fun in pitting a grandmaster against a novice at chess, where randomness plays little part. Of course, in the longer run, expertise steadily wins out, so that, for example, good poker players can live off their skill. In the real world, the visibility of random processes is often obscured by other factors, so to begin to understand randomness we start with games.

Let's use the example of flipping or tossing coins. This is a random process. The coin may land on either *heads* or *tails*. In the real world, flipping is not totally random. With a little practice and judicious placing of the coin in your hand, it is possible to flip it in such a way that it usually lands on the side you want. However, for the sake of our exercise here, let's imagine a flipping machine that flips a fair coin in a consistent but totally unbiased way. Each flip of the coin is a *trial*. Each repeated flip of the coin is an *identical* trial – the conditions do not change from flip to flip, and each trial is *independent* – that is to say, no result of any flip can have any impact on the result of any other one. I know for a fact that, unlike human beings, coins are unconscious and have no memory. Thus, the fact that they have landed on *heads* or *tails* in the past, or on the most recent throw, can have no possible bearing on how they will land the next time.

The coin-tossing example also has some historical interest. John Kerrich, a British mathematician, found himself interned in Denmark when it was invaded by the Nazis in 1940. He passed some of the time tossing a coin and recording the results – 10,000 times. Now, we can simulate Kerrich's experiment on any laptop in a few seconds, but Kerrich's attempt, long before the advent of powerful computers, was an early empirical confirmation of the law of large numbers.

We really use the idea of a coin flip to imagine any random process defined by having only two equally probable outcomes. It is the mathematics that is important, not the attempt to approximate them with a physical process. The empirical nature of coin flipping has been studied intensively by Persi Diaconis, Professor of Statistics and Mathematics at Stanford University, who showed that coins flipped manually are marginally more likely to land on the face that was uppermost before the flip. Two Berkeley students tested his results, empirically performing 40,000 coin flips, taking 1 hour a day for an entire semester. Diaconis was a former professional magician who paid his way through college by playing poker, demonstrating the continued strong link between games of chance and statistics that started with the exchange of letters between Blaise Pascal and Pierre de Fermat about a game of dice that kicked off the whole modern mathematical analysis of probability.

We can summarise the law of large numbers in the equation below. The underlying probability of any outcome within a trial sample space equals the proportion of times that outcome occurs. The proportion will approach a limit, which defines its probability:

$$P = \frac{\text{Frequency of outcome}}{\text{Number of trials}}$$

In our case, we already know 'in theory' that our coin is fair so that the proportion of *heads* should equal the proportion of *tails*, and both should approach one half.

Bernoulli trials

Flipping a coin is an example of a *Bernoulli trial* named after the mathematician Jacob Bernoulli who set out the mathematics of such trials in *Ars Conjectandi*, one of the earliest works on probability, in 1713. Almost anything that can happen in the natural and social world can be modelled on such a trial if we divide the trial outcomes into (1) one that interests us (sometimes called *success*) and (2) every other outcome: someone is promoted or not, smokes marijuana or not, is arrested or not, dies or survives, your computer crashes or not, an organism reproduces itself or not and so on. All these examples reduce to the following:

An outcome of interest (*success*).

Any other outcome.

Let's repeat Kerrich's experiment 10 times using Excel (details are in the appendix to this chapter if you would like to try this at home). Let 1 stand for *heads* and 0 for *tails*. I am interested in two main quantities. The first is the difference between the number of *heads* I would *expect* to obtain and the number of *heads I observe*. We know that the total for *heads* should be equal to the total number of trials multiplied by 0.5. This is because we think the coin is fair, and the probability of landing on *heads* or *tails* ought to be equal. However, it would be very unusual indeed if our coin landed *heads* on one flip followed by *tails* on the next one.[1] Thus, there will be some difference between the number of expected *heads* and the number of *heads* we actually *observe* at any one point in the series of trials. We can call the total we would expect the *expected value*. If we take the difference between the running total of *heads* we *observe* and this *expected* value, we can call this difference the *error*. In this context, error does *not* mean mistake or fault. There is nothing wrong with our experimental set-up, nor is it in any way unusual to see this error. On the contrary, it would be the absence of this error that would be unusual and call for some explanation. Rather *error* is the term we use to describe the impact of randomness. We can now divide our results between the *expected value*, the *observed value* and the *error*. We can also keep a running total of the *estimate* that we obtain for P(*heads*) as our trials progresses, by dividing these results by the total number of trials so far.

$$P(heads) = \frac{N \text{ heads observed}}{N \text{ trials}}$$

N observed *heads* = (N expected *heads* − Error)

The law of large numbers describes what happens as the number of trials grows. For a small number of trials, the *expected* value, the *observed* value and the *error* are all small in absolute terms. If we take the average across all trials, the error is large and our estimate of P(*heads*) is poor. Indeed, every time we have completed an odd number of trials, there *must* be some error, because the expected value is not a whole number, while our running total of *heads* can only increase by one at a time. Table 2.1 (generated using the random number function in Excel) shows how a series of trials

---

[1]We can easily calculate just how unusual it would be. Each flip of the coin carries a probability of one half of landing on a particular face, so a run of *any* particular sequence of faces of length N, including HTHTHTHT … must be $1/2^N$. The chance of a sequence of five alternating pairs of heads and tails is less than 1 in 1000 ($1/2^{10} = 1/1024$).

might build up. First came a 'tail', followed by another and another, the fourth flip produced the first head, then the fifth a tail again and so on. As the number of trials increases, however, the total error continues to grow erratically, but by much less than the number of trials. After only 100 trials, our estimate of P is 0.51. By 10,000 trials, it is 0.506.

Table 2.1 Ten thousand coin flips

Number of Trials	Sequence of Results	Observed Value	Expected Value	Absolute Error	Estimate of $P(\hat{P})$	$\hat{P} - P$
1	T	0	0.5	−0.5	0.00	−0.5000
2	TT	0	1	−1	0.00	−0.5000
3	TTT	0	1.5	−1.5	0.00	−0.5000
4	TTTH	1	2	−1	0.25	−0.2500
5	TTTHT	1	2.5	−1.5	0.20	−0.3000
6	TTTHTH	2	3	−1	0.33	−0.1667
7	TTTHTHH	3	3.5	−0.5	0.43	−0.0714
8	TTTHTHHT	3	4	−1	0.38	−0.1250
9	TTTHTHHTH	4	4.5	−0.5	0.44	−0.0556
10	TTTHTHHTHH	5	5	0	0.50	+0.0000
50		28	25	3	0.56	+0.0600
100		51	50	1	0.51	+0.0100
500		254	250	4	0.51	+0.0080
1000		514	500	14	0.51	+0.0140
10,000		5057	5000	57	0.5057	+0.0057

Figure 2.4 shows the results of 10 repetitions of a series of 10,000 flips. Each different-shaded line shows the value for the estimate of P after each of the 10,000 flips within the series. In every case, within a few hundred flips, the error is squeezed, and the estimates approach the true underlying probability. Figure 2.5, which plots only the first 500 flips in each series, shows clearly that the way in which the error reduces is *random*. It is neither a steady nor smooth process, but it does progress with an inexorable force.

Now look at Figure 2.6. It records the *total* error – that is, the difference between the *observed* and the *expected* total number of *heads*. The error *increases* with the number of trials, but not in any predictable way. It can take the form of an increase over time in the surfeit of *heads*, or in the deficit of *heads*, or in a mixture of both that oscillates between the two. It is genuinely random. If it *could* be predicted, then casinos would no longer make money. There is *absolutely* no tendency for any excess of *heads* or *tails* to 'correct' itself. In fact, irregular long strings of *heads* or *tails* is exactly what we *should* expect. To think otherwise, is to imbue coins with memory and a conscience,

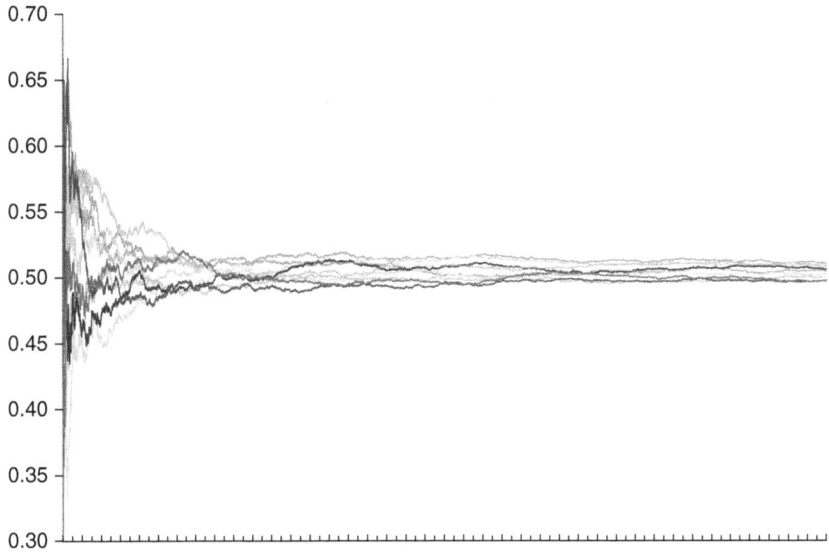

Figure 2.4 Estimate of P(heads) over 10,000 trials, 10 repetitions

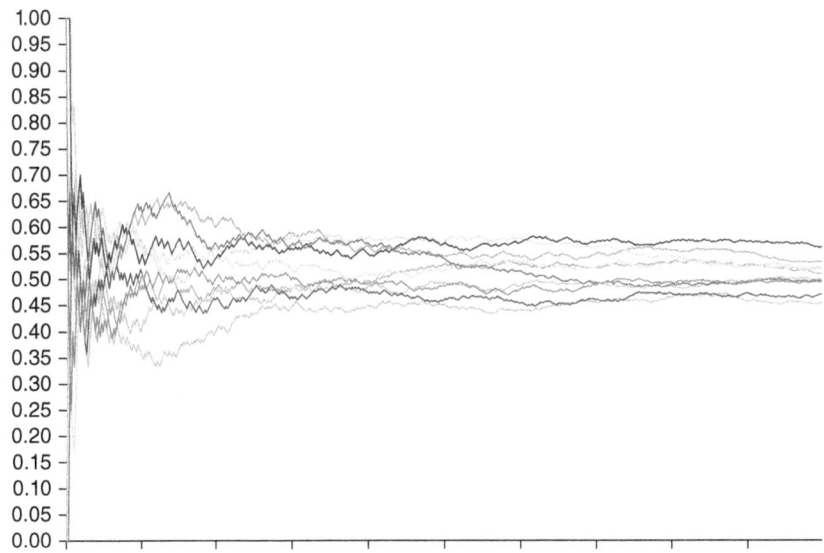

Figure 2.5 Estimate of P(heads) over 10,000 trials, 10 repetitions, first 500 flips

and it is known as the *gambler's fallacy* for fairly obvious reasons. In a game of chance that paid out when our coin landed *heads* up, and we treat these repetitions as the history of 10 players, 6 would have made a profit and 3 a loss. However, it would have been impossible to determine their future performance from their early positions, and there is no tendency for a 'run of bad luck' to be followed by better fortune. Randomness really is random.

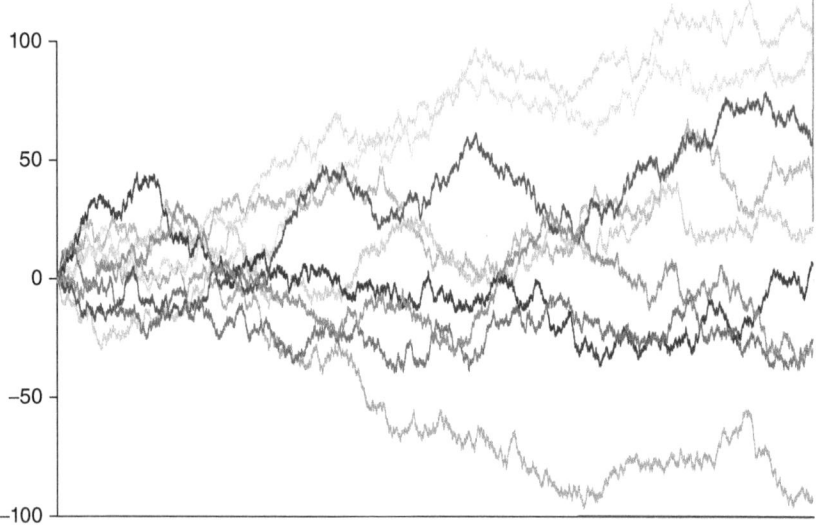

Figure 2.6 Total error, 10 repetitions, 10,000 flips

It cannot be stressed too much that there is absolutely no evidence of the process 'balancing' or 'evening out'. *Over time, error increases.* However, as the number of trials increase, we move inexorably towards a better estimate of our underlying probability because *the total error increases more slowly than the number of trials*. In fact it turns out that for a Bernoulli trial with equal probability of each outcome, the absolute error will be around

$$0.5 * \sqrt{\text{Number of trials}}$$

This does *not* mean that the absolute error in any one series of trials will equal this amount. This too is a limit to which the error would converge were we to run a very large number of repetitions of our run of 10,000 trials. The formula suggests that we should expect an absolute error around 50.

$$\frac{1}{2} * \sqrt{10,000} = \frac{1}{2} * 100 = 50$$

Our results came out at

−93	+42	+105	+92	+22	+80	+7	+58	−25	−29

Ignoring the sign, this is an average error across all 10 repetitions of 55, well within the range that we might expect.

The law of large numbers can be proven mathematically (which would take us beyond the scope of this book), but its essence is remarkably simple: individual random events are *genuinely* unpredictable. We could spend the rest of our lives study-ing patterns of random coin flips but still be not one jot wiser about whether the next result will be *heads* or *tails*. However, the outcomes of large collections of random events can not only be predicted but also predicted with tremendous precision. The ability to do this has nothing whatsoever to do with things 'balancing out'. Rather it is a feature of the way in which we can move backwards and forwards between proportion and probability once we can repeat even modestly large numbers of trials. The proportion of times an outcome within the sample space occurs equals its probability, and vice versa.

This may, at first sight, seem to be an underwhelming discovery, and moreover, a rather old one. Jacob Bernoulli worked out most of it three centuries ago. However, it was only about 100 years ago that it came to be realised that it could resolve an other-wise intractable problem with sampling. Before we see how, we first need to complete our understanding of probability.

The probability distribution

We have seen that any individual outcome of large numbers of repeated trials tends towards a limit that represents the underlying probability of that outcome, expressed as a proportion between 0 and 1. It follows from our definition of a trial, that the sum of these proportions for each outcome in the sample space *must be exactly one*. Something must happen in each trial. The total of all the outcomes must therefore equal the total number of trials. The relative *frequency* of each outcome in the sample space in any series of trials will therefore be the proportion or times they occur and equal to their probability. We can therefore display the outcomes of a trial that has been repeated as a *distribution* showing the *frequency* of each outcome in the sample space. We can produce a *frequency table* if outcomes are discrete or *summary statistics* (like the mean, median, standard deviation or IQR) if the outcome is continuous. We can also represent this distribution visually using *bar charts* or *histograms*.

Table 2.2 is a frequency table based on results from the eighth round of the European Social Survey showing the sex of respondents. There were 52,142 identical repeated independent trials comprising the question about the sex of the respondent. Of them, 24,948 respondents said that they were male, which is 0.478 of all the trials. We could describe the probability of that outcome as $p = 0.478$. This does not mean than that anyone in the survey was 47.8% male. Our probability refers to the series of all the repeated trials, not the result of any individual trial. It *does* mean that if we selected anyone *at random* from all the respondents, we would have a *long-run* prob-ability of 47.8% of picking a man. We would have a $p = 0.522$ of picking a woman. And because these two outcomes described the complete sample space, we would have $0.478 + 0.522 = 1 = $ the absolute certainty of picking either a man or a woman.

Table 2.2 Sex of respondent

Sex	Frequency	Percentage	Proportion
Male	24,948	47.8	0.478
Female	27,194	52.2	0.522
Total	52,142	100	1

Note. European Social Survey Round 8.

What does 'at random' mean in this context? It would have to mean that any of the respondents was as likely to be picked as any other – that is, they would have to have equal probabilities of selection. The elaborate contraptions of urns, balls, electronic random number generation equipment and so on that gets used for lotteries of various kinds works on this principle. Unless every ticket had exactly the same chance, who would buy a ticket with a poor one!

Table 2.3 shows respondents' (dis)agreement with a statement about homosexuality. Note that there is a small probability (about $p = 0.04$) that respondents either refuse to answer the question or say they don't know what their answer is. This is probably a small enough proportion for us to disregard these answers. Even if every respondent in this situation could really be placed in one of the other categories, if some method were to be devised for ethically discerning their 'true' response, it would have little impact on the rest of the table. We can treat these responses as 'missing at random' and discard them from our analysis and recalculate our probabilities using only the 50,250 respondents who *did* give a definite response. We could sum the proportions who agree or agree strongly and see that $p = 0.665$ of respondents chose one of those two options. This proportion would vary by country within the survey. Our facility with moving between proportions and probabilities is shown by the way we would almost certainly describe this: either as 'proportionately more people' accepting homosexuality in some countries compared with others or people being 'more likely to' accept homosexuality. Yet another way of putting this would be to say that the probability of accepting homosexuality was *conditional upon*, or depended upon, the country in which respondents were interviewed. We will look at conditional probability a little later.

Table 2.3 Response to the statement 'Gays and lesbians should be free to live life as they wish'

Response	N	%	Proportion	Adjusted Proportion
Agree strongly	17,589	33.7	0.337	0.350
Agree	15,843	30.4	0.304	0.315
Neither agree nor disagree	5713	11.0	0.110	0.114
Disagree	4658	8.9	0.089	0.093

Response	N	%	Proportion	Adjusted Proportion
Disagree strongly	6447	12.4	0.124	0.128
Refusal	571	1.1	0.011	-
Don't know	1317	2.5	0.025	-
No answer	8	0.0	0.000	-
Total	52,146	100	1	1

Note. European Social Survey Round 8.

Table 2.4 is the frequency table for the weight of 1827 babies that were part of the 1958 National Child Development Study (NCDS). Because weights vary almost continuously, they have been 'binned' into intervals of 5 ounces at a time. Again, if I started picking babies at random, the probability of getting a baby in any of these weight intervals would be the same as the proportion of all the 1827 babies that were found in that interval. I could also add intervals together. A quick look at the table shows that almost half of all babies have a birthweight between 110 and 129 ounces: the probability of falling within that range is about $0.116 + 0.107 + 0.135 + 0.102 = 0.46$. I could also describe the probability distribution shown in the frequency table by using the *mean* (118 ounces) and the *standard deviation* (18.4 ounces). That would give me a good idea of what size babies tended to be (not 10 ounces or 1000 ounces) and how closely they clustered around this average (fairly closely). The median baby weight would be the baby ranked $\frac{1827+1}{2} = 914$ in ascending order, which comes in the interval 115 to 119 ounces.

Table 2.4 Birthweight of babies in ounces

Weight in Ounces	%	P	N
up to 54	0.1	0.001	1
55–59	0.2	0.002	4
60–64	0.3	0.003	5
65–69	0.2	0.002	3
70–74	0.6	0.006	11
75–79	0.8	0.008	15
80–84	1.4	0.014	25
85–89	2.5	0.025	45
90–94	3.0	0.030	55
95–99	6.0	0.060	110
100–104	7.9	0.079	145
105–109	7.5	0.075	137

(Continued)

Table 2.4 (Continued)

Weight in Ounces	%	P	N
110–114	11.6	**0.116**	212
115–119	10.7	**0.107**	196
120–124	13.5	**0.135**	247
125–129	10.2	**0.102**	186
130–134	5.5	0.055	100
135–139	5.6	0.056	102
140–144	6.1	0.061	112
145–149	2.1	0.021	39
150–154	1.8	0.018	33
155–159	0.8	0.008	15
160–164	0.8	0.008	15
165–169	0.4	0.004	8
170–174	0.2	0.002	3
175–179	0.1	0.001	2
185–189	0.1	0.001	1
Total	100.0	1.000	**1827**

Note. National Child Development Study teaching data set.

Bar charts and histograms

Probability distributions can be displayed visually using the following rule. The proportion of the total area of the chart that represents each outcome is the same as the probability of that outcome in the sample space. For bar charts, as long as the bars are of equal width, this means that their height is proportional to the probability, or to the frequency (since the frequency of each outcome divided by the total number of trials is the same as the probability). Figure 2.7 shows our example of attitudes to homosexuality, with the refusals and don't knows excluded. At a glance, we can see that most people agree with the statement.

Histograms can display *continuous* as well as *discrete* probability distributions. In a histogram, instead of separate bars for each outcome, the *area* above a range of outcomes represents that interval's share of all the outcomes recorded and thus its probability. Figure 2.8 displays the probability distribution from Table 2.4. Its shape may appear familiar. Most babies have weights near the mean. As we move farther from the mean, the number of babies declines, until at the extremes or 'tails' of the distribution (shown in black) there are very few babies indeed who are either very small or very large. Patterns similar to this one occur often in nature. It is referred to as a *'normal'*, *'Gaussian'* or *'bell'* curve, and it can result when many different factors

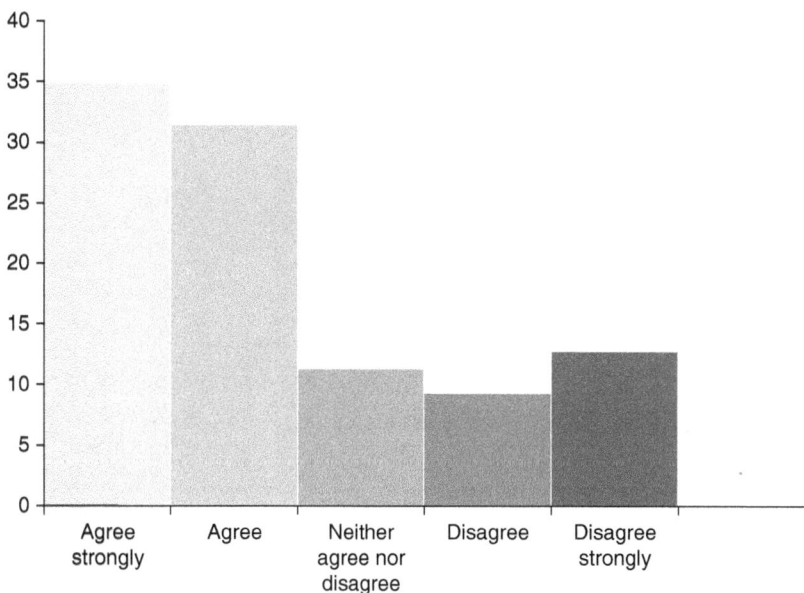

Figure 2.7 Response to the statement 'Gays and lesbians should be free to live life as they wish'

Note. European Social Survey Round 8.

influence the magnitude of a variable, as the second part of the appendix to this chapter demonstrates. It does *not* mean that whenever there are many factors, we always get such a distribution. Nor does it mean that empirically occurring distributions that are *similar* to a Gaussian curve are *identical* to it. As we shall see later, this can be a crucial distinction to keep in mind. However, normal and approximately normal curves have two extremely useful properties that we will exploit.

The first is that they can be described by just two values: the *mean* and the *standard deviation*. The second is that because of this property, we can always predict the area under the curve in a histogram that lies to the right or the left of any values of a variable that they describe. In a perfectly normal distribution, about 68% of the observations are within 1 *SD* above or below the mean. About 95% of them lie within approximately *2 SDs* (actually 1.96). Instead of using the original measurement scale for a variable, we can translate it into standard deviation units called z-scores. You will sometimes see the formula for a z-score presented as:

$$z = \frac{X - \mu}{\sigma}$$

The Greek letter pronounced 'myoo' is a symbol often used for the mean, while 'sigma' is used for the standard deviation. Thus, the formula says that the z-score

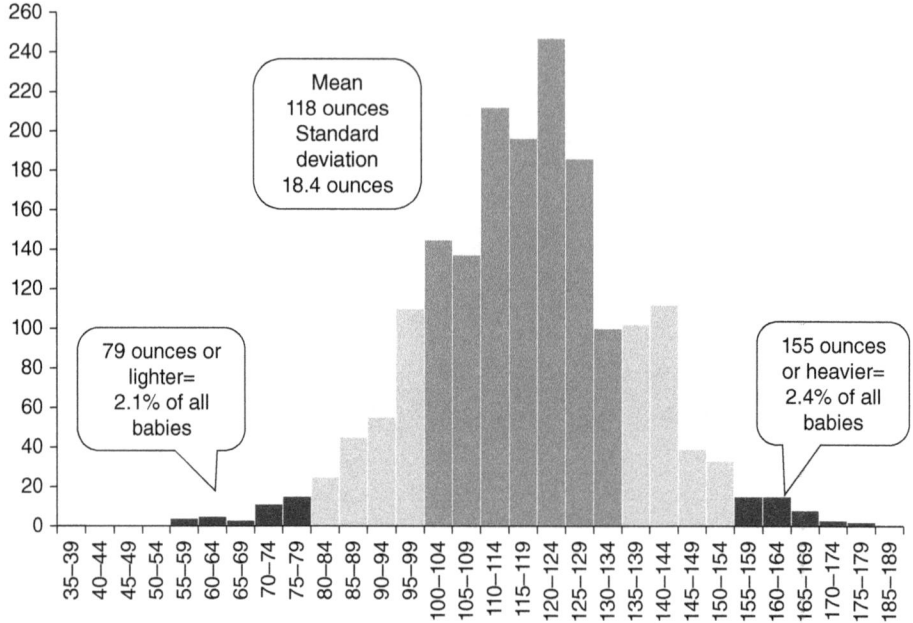

Figure 2.8 Birthweight of babies

Note. National Child Development Study teaching data set.

is obtained by subtracting the mean from any value the original variable takes and dividing the result by the standard deviation.

Thus, in our example of babies' weights, we can calculate their mean and standard deviation, which come out at a mean of 118 ounces and a standard deviation of 18.4 ounces. If the distribution of babies' weights were perfectly normal, then we'd expect about 68% of the babies to have a weight between 118 − 18.4 = 99.6 ounces and 118 + 18.4 = 136.4 ounces. We'd expect 95% to have a weight between about 118 − (1.96 * 18.4) = 81.9 and 118 + (1.96 * 18.4) = 154.1 ounces. That means that we'd expect about 2.5% of babies to weigh less than 82 ounces and 2.5% to weigh more than 154 ounces. If you study Figure 2.8, you will see that the distribution fits a normal curve not too badly!

Heights of Union soldiers

Figure 2.9 shows the height of more than 20,000 Union soldiers at the time of the Civil War. The area in the histogram covering the interval of heights from 65.0 inches to just below 67.0 inches has been shaded dark grey. Of them 5578 soldiers were in this height range – that is, 27.6% of all the soldiers in the sample. The green area occupies 27.6% of the area of the histogram. The probability of any soldier chosen at random being within this height interval would be 0.276. This is another empirical distribution, but it conforms very closely to what a Gaussian distribution would lead us to expect. The survey was organised by Benjamin Gould, one of the founders of the US National Academy of Science in 1863, an astronomer, and amongst his many other scientific roles, actuary to the US Sanitary Commission, who in the face of often-determined bureaucratic resistance collected and tabulated the vital statistics of troops in order to test 'those hygienic and physiological laws which are already known' and 'to discover and apply such other laws as might affect the welfare and success of our soldiers' (Comstock, 1922, p. 162). The data he compiled proved an immensely valuable resource in the study of health, demography and many other fields. Soldiers' heights were measured to the nearest half inch. The mean height of the 20,207 soldiers he studied was 67 inches and the standard deviation found to be 2.58 inches. We might therefore *expect* to see a total of around 0.05 * 20207 ≈ 1010 soldiers who were either *taller* than 67 + 1.96 * 2.58 = 72 inches or shorter than 67 − 1.96 * 2.58 = 62 inches. Gould *observed* 1055 such soldiers, remarkably close to what we might expect if height followed a perfectly Gaussian distribution.

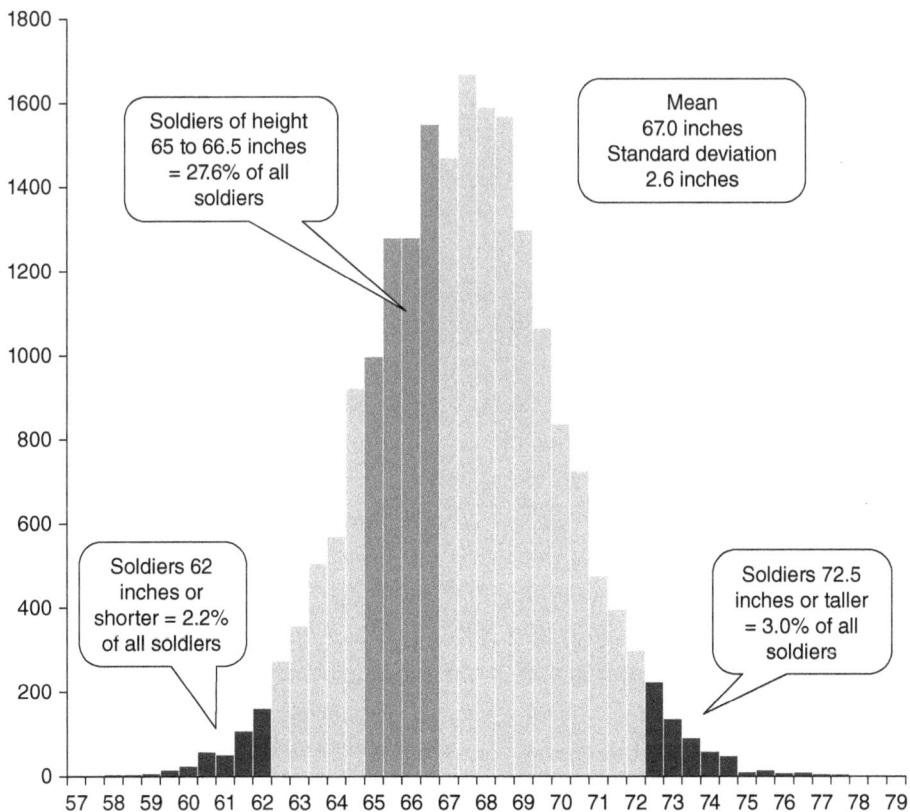

Figure 2.9 Heights of Union soldiers

Note. Data downloaded from www.nber.org/gould/ and described in Costa (2004).

The probability distributions we've examined of coin flips, sex, attitudes to homo-sexuality, babies' weights and soldiers' heights are *empirical*. They were based on trials that have actually been conducted and outcomes recorded. But we have also imagined *theoretical* probability distributions of what we'd expect to happen, or think logically must happen, given certain conditions: *if* a long-run probability of a coin landing *heads* or *tails* was 0.5, or *if* the distribution of babies' weights or soldiers' heights was approximately normally distributed. Comparing what we actually do *observe* and what we could *expect* to observe, *if* we assume that some condition is met, is a procedure that proves immensely useful in inference, as we shall see in the next chapter.

Finally, remember that the empirical results we have examined here as probability distributions are all just as much about randomness and probability as our coin flips. *Before* the data was collected, we could not know what the results might be! The processes generating the data were like tossing a coin. Just as we could never understand and measure all the forces that might determine the outcome of an individual toss of a coin, neither could we explain every soldier's height or every babies' weight, or every person's views about homosexuality. Once we capture the probability distribution for these variables, however, we can then investigate what might affect it.

Probability distributions and variables

By now, it should have become clear that probability distributions and variables are the same beast. Each variable describes a trial with a sample space. The values of variables correspond to the trial outcomes in the sample space for that variable. Each observation or case is one instance of that trial. This raises the question 'How do we ensure that the trials are identical and independent?' If they were not identical and independent, there would be no point in their repetition because we would have no means either of comparing them or of calculating the results obtained from more than one trial. It would be rather like trying to add apples and bicycles together. If I flip a coin a few times, I can reasonably claim to be repeating an identical and independent trial. But in what way is asking a collection of *different* people a survey question the repetition of an *identical* and *independent* trial? I must be able to claim that the respondents share something *in common* that gives them the same status as the virtual coin I flipped, and that it is this that I can make probability statements about, just as I could about the coin. What they have in common is *membership of a target population*. Just as my probability statements about coin flips refer to the coin, and not to any individual outcome from a flip, so too do my statements about survey respondents, whether expressed as proportions or probabilities, refer to them as a group and not to individual members of that group. The target population can be

48% male. My probability of picking a man at random from it can be 0.48, but there are no individuals in it who are 48% male, or 67% agree with a statement or possess precisely the mean age, income or height of the target population. Neither are there individual soldiers who have 2.5% of their height above 6 feet, or babies whose individual weight has any standard deviation.

Conditional probability with categorical variables

Like probabilities, we also use *conditional* probabilities all the time without necessarily thinking about it. In fact every time we make some kind of association we implicitly use a conditional probability. I see an ageing customised car accelerating noisily away and may think 'boy racer' even if I have not seen the driver. Why? Because older drivers typically drive more expensive cars and more sedately, customising cars is something done far more often by men. The logic behind this is that the probability of being a young male as opposed to a young female or older driver is *conditional upon* the type of car being driven. Conditional probabilities are usually shown using contingency tables. Earlier we saw the probability that adults in Europe accepted homosexuality as a normal way of life. Table 2.5 shows the same results broken down by a selection of countries. To keep things simple, I have excluded those who said they 'neither agreed nor disagreed' and divided the remainder into agree and disagree, whether or not they did so 'strongly'. This table therefore shows the results of *two* series of repeated trials: one recording respondents' reaction to the statement and the other the country in which they lived.

Table 2.5 Response to the statement 'Gays and lesbians should be free to live life as they wish' by country, 2016

Country	Agree (%)		Disagree (%)		n
Netherlands	97.7	1596	2.3	38	1634
Spain	95.5	1704	4.5	81	1785
United Kingdom	94.7	1691	5.3	95	1786
Germany	93.8	2482	6.2	165	2647
Ireland	92.3	2327	7.7	195	2522
Italy	81.9	1685	18.1	373	2058
Estonia	71.0	1125	29.0	460	1585
Hungary	48.2	550	51.8	592	1142
Lithuania	33.6	464	66.4	918	1382
Russian Federation	18.0	332	82.0	1510	1842
All	75.9	13,956	24.1	4427	18,383

Note. European Social Survey Round 8.

If you study the table, you can see a number of features. Each compartment of the table containing a number is referred to as a 'cell'. The lightly shaded cells corresponding to each combination of an individual row variable and a column variable category are referred to as the *body* of the table. The darker shaded final column and row comprise the *margins* of the table. The contingency table comprises two frequency tables, or probability distributions, set side by side. The third column shows the frequency of 'agree' responses, and the fifth column the frequency of 'disagree' responses. Look closer still and you will see another two frequency tables. The sixth column shows the frequency of response by country, regardless of view of homosexuality, while the final row shows the frequency of response by view of homosexuality, regardless of the country of the respondent. In order to make the table clearer, the results within each row of the table have been standardised by presenting them as a percentage. Thus, for example, in Spain 81/1785 = 4.5% of respondents said that they disagreed with the statement, while in Hungary, 592/1142 = 51.8% of respondents did so. Standardising in this way makes it easier to compare different countries. If we ask whether views about homosexuality are *contingent upon*, or *conditional upon*, or *conditioned by*, or *depend upon* or *associated with* country, how do we answer that question? We would look down the columns to see whether the row percentages change. If they do, then we have evidence that views about homosexuality vary by country or that the probability distribution for views is conditional upon that for country. On the contrary, if the percentages varied little from one country to another, we could say that the two probability distributions were not associated, or that they were independent of each other. Let's examine this last point in greater detail.

Table 2.6 shows some more results from the 1958 NCDS survey of babies, recording the occupational social class of the babies' fathers and the smoking behaviour of the babies' mothers. First note that although the original survey questions were directed at each of the babies' parents, the unit of analysis is really the baby, so that we have *two* sets of trials directed at *the same* subject, just in the same way as we might have recorded any other two variables about the baby, such as its height or weight at birth. Table 2.6 is a contingency table because it shows the probability of the mother smoking *contingent upon* the occupational social class of the father. Conditional probabilities can be written using a long vertical bar, which is read as 'conditional upon', so that our table shows *four* probability distributions:

1   P(class): The marginal probability of being in each class, shown in the right-hand margin of the table. P(non-manual) = 468/1739 = 0.28; P(manual) = 1253/1739 = 0.72.
2   P(smoking): The marginal probability of smoking shown in the bottom margin of the table. P(does smoke) = 550/1739 = 0.32; P(does not smoke) = 1189/1739 = 0.68.

3 P(smoking | non-manual): The probability of smoking conditional on being non-manual class. P(smokes | non-manual) = 112/486 = 0.23; P(does not smoke | non-manual) = 374/486 = 0.77.

4 P(smoking | manual): The probability of smoking conditional on being manual class. P(smokes | manual) = 438/1253 = 0.35; P(does not smoke | manual) = 815/1253 = 0.65.

Is the probability of the mother smoking conditional on the social class of her partner? A quick answer to this question is provided by comparing the marginal and conditional probability distributions. If they are *not* the same, then we have evidence of an association between the two marginal probability distributions. It looks as if there is such an association, although it is not very strong: around one quarter of the female partners of non-manual fathers smoked, compared to about one third of the female partners of fathers who worked in manual jobs.

Table 2.6 Father's class by whether mother smokes

Class	Whether Smokes		
	Yes	No	All
Non-manual	112	374	486
	0.23	0.77	0.28
Manual	438	815	1253
	0.35	0.65	0.72
All	550	1189	1739
	0.32	0.68	1.0

Note. National Child Development Study teaching data set.

A more formal way in which we could answer this question would be to see whether the two sets of trials were *independent* of each other: that is, just like tossing a coin twice, the result of one trial has no impact on the result of the other one. We can work out what results we would *expect* to see if this was the case by using the two frequency distributions in the margins of the table. If we multiply the observed frequencies in the table margins together, and divide by the total number of observations, we will get the number of cases in each cell of the table that we would *expect* to see *if* the two distributions were indeed independent. Table 2.7 shows the calculations. You can check the numbers in each cell and see that they preserve the distribution in the table margins unchanged. Table 2.8 presents the same table but with column percentages (in bold) and row percentages (in italics) shown, so that you can check that with our expected frequencies:

- The proportion of mothers who smoke is the same for babies with fathers in manual and non-manual occupations.
- The proportion of manual fathers is the same for others who do and do not smoke.

Table 2.7 Father's class by whether mother smokes: *expected* frequencies

	Whether Smokes		
Class	Yes	No	All
Non-manual	=550*486/1739 = 153.7	=1189*486/1739 = 332.3	486
Manual	=550*1253/1739 = 396.3	=1189*1253/1739 = 856.7	1253
All	550	1189	1739

Note. National Child Development Study teaching data set.

Table 2.8 Father's class by whether mother smokes: *expected* frequencies with column (in bold) and row (in italics) percentages

	Whether Smokes		
Class	Yes	No	All
Non-manual	**27.9** *31.6* 154	**27.9** *68.4* 332	486
Manual	**72.1** *31.6* 396	**72.1** *68.4* 857	1253
All	550	1189	1739

Note. National Child Development Study teaching data set.

There is clearly a difference between the frequencies we *observed* in Table 2.6 and those we would *expect* if the distributions were independent. This raises two questions. Our data is from a sample of 1739 babies, not every baby born in 1958. Would it be safe to generalise our results to that population? We look at this in Chapter 3. The difference between our *observed* and *expected* frequencies is pretty modest. How small would it have to get before we could conclude that the difference did not matter? We look at this in Chapter 4.

There are two further things to note about our contingency table. Conditional probabilities are *not* symmetrical. That is to say, the probability of a mother smoking, given that the father has a manual occupation is *not* the same as the probability of a father having a manual occupation, given that the mother smokes. The first probability is described by the *row percentages* in Table 2.9. The second probability is described by *the column percentages*.

From Table 2.9, we can see that

P(mother smokes | manual class father) = 0.350

P(manual class father | mother smokes) = 0.796

Table 2.9 Father's class by whether mother smokes: *observed* frequencies with column (in bold) and row (in italics) percentages

| Class | Whether Smokes | | |
	Yes	No	All
Non-manual	**20.4**	**31.5**	486
	23.0	*77.0*	
	112	374	
Manual	**79.6**	**68.5**	1253
	35.0	*65.0*	
	438	815	
All	550	1189	1739

Note. National Child Development Study teaching data set.

You may come across the following rather intimidating looking formula for conditional probability:

$$P(A|B) = \frac{P(A \text{ and } B)}{P(B)}$$

In the everyday language of events, it says, 'Considering only what happens if B is true, what is the probability of A being true?' Or using the language of processes and proportions 'What proportion of all Bs are also As?' For our table, we could express it as:

$$P(\text{Mother smokes} | \text{Father's class}) = \frac{P(\text{Mother smokes and Father's class})}{P(\text{Father's class})}$$

$$P(\text{Mother smokes} | \text{Father's class}) = \frac{P(438/1739)}{P(1253/1739)} = \frac{438}{1253} = 0.35$$

Study the table, and you'll see that the column percentages, or conditional probabilities, in each cell all correspond to this formula, whether you use the raw numbers or the percentages.

Our brains are not well wired to deal directly with probabilities like this. Much easier is to think in terms of frequencies. We could take the numbers in Table 2.9 and produce a new version of the table that described the situation for 100 babies, shown in Table 2.10. We could then summarise the table much more simply as:

Out of every 100 babies, 32 have mothers who smoke. For those with fathers in manual jobs, this rises to 35 (25.2/80.1), and for fathers in non-manual jobs, it falls to 23 (6.4/27.9).

Table 2.10 Father's class by whether mother smokes: *observed* frequencies per 100 cases

Class	Whether Smokes		
	Yes	No	All
Non-manual	6.4	21.5	27.9
Manual	25.2	46.9	80.1
All	31.6	68.4	1739

Note. National Child Development Study teaching data set.

The second key point to note is that conditional probabilities are not in themselves evidence of *causation*. From the evidence in Table 2.9, it would be just as mistaken to jump to the conclusion that mothers smoke because fathers have manual occupations as it would be to think that fathers had manual jobs because mothers smoked!

Conditional probability with continuous variables

What about examining independence and conditional probability with continuous distributions? Here, we can use *scatter plots* instead of contingency tables. A histogram plots the number of cases (on the vertical axis) against the values for the probability distribution from a trial on the horizontal X axis. A scatter plot plots both the outcomes from a series of a pair of trials as *coordinates*, with the values for one outcome on the horizontal axis and the values from the other on the vertical one. Figure 2.10 plots height in centimetres of the NCDS girls at age 11 against their weight in kilograms. The coordinate with the solid black centre in Figure 2.10 represents a girl who was 131 cm tall and weighed just under 50 kg. If you study the scatter plot, you can see that the coordinates form a definite pattern. They are not scattered randomly about the graph but tend to form a rough ellipse that stretches from the bottom left of the graph to the top right. We could summarise the shape of the ellipse by a straight line running through its centre, which is shown as a dotted line. There were no short children who were heavy, and very few tall girls who were light, so that the top left and the bottom right quadrants of the graph are relatively empty.

We would expect there to be some association between height and weight. Someone who is taller is also likelier to be heavier, but we would not expect the association to be perfect: there are short and plump as well as tall and skinny people. By contrast, we would not expect there to be much, if any, relationship between height and ability at maths. The NCDS children took a maths test at age 11. Figure 2.11 shows the results for girls. The coordinates are scattered fairly randomly over the graphic. We could safely conclude that the trials for height and maths ability were independent. Being good at maths does not make you tall, nor does height prepare you for a maths test!

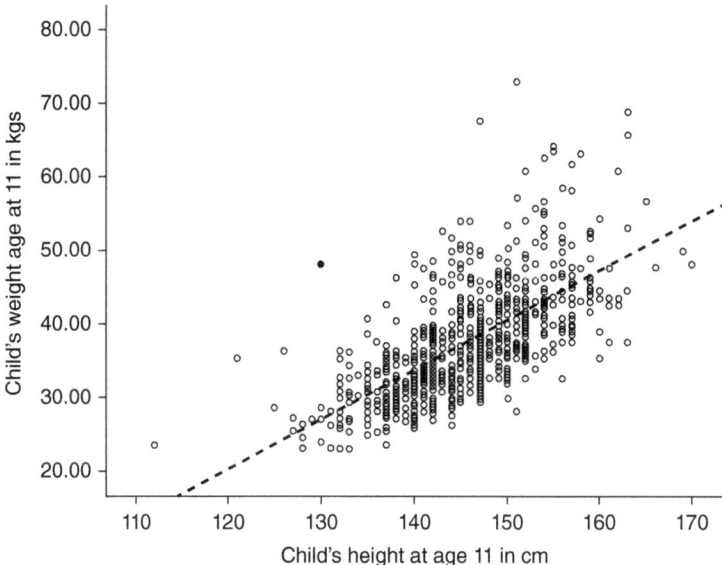

Figure 2.10   Height and weight at age 11, girls

Note. National Child Development Study teaching data set.

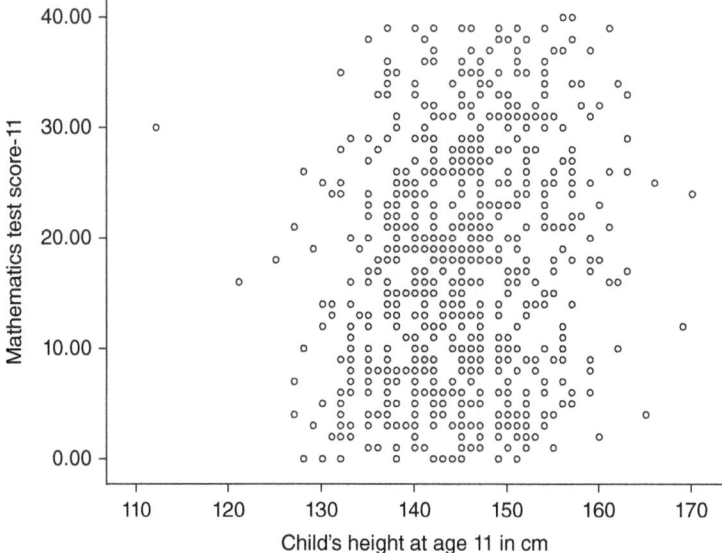

Figure 2.11   NCDS girls: height by maths ability at age 11

Note. NCDS = National Child Development Study. Data from NCDS teaching data set.

Pearson's *r*

The *Pearson product moment correlation coefficient r* is one way of summarising the tendency of coordinates to cluster along a straight line like this. To understand how it is calculated, let's think of redescribing the values of each of our variables. For each case, we could describe their height as the difference between their height and the *mean value for height for all the* cases in our data. This gives us the value of the *residual* from the mean for each case. Since our height variable was on the vertical Y axis in our scatter plots, we can represent it with the letter 'y' for short. For each case, we could thus calculate

$y - \bar{y}$ [where $\bar{y}$ stands for 'the mean of y' and is read 'y-bar']

(Remember that this is what we did when calculating a standard deviation.)

We could do the same for our weight variable using the letter 'x' to stand for the weight variable:

$x - \bar{x}$ [where $\bar{x}$ stands for 'the mean of x']

Now if height and weight were associated, we'd expect most people who were above average (mean) weight to *also* be above average (mean) height. That's just a way of saying that we'd expect many cases in the top right corner of our scatter plot. In exactly the same way, we'd expect most people who were below average (mean) weight to *also* be below average (mean) height. That's just a way of saying that we'd expect many cases in the bottom left corner of our scatter plot.

What happens if we multiply the x and y residuals together: $(y - \bar{y}) * (x - \bar{x})$?

- For cases above the mean on height *and* above the mean on weight: We have a positive number times another positive number; this produces a positive number that increases in size the taller or heavier the case is.
- For cases below the mean on height *and* below the mean on weight: We have a negative times another negative number; this produces a positive number that increases in size the shorter or lighter the case is.
- For cases below the mean on height *but* above the mean on weight: We have a negative times a positive number; this produces a negative number that increases in size the shorter or heavier the case is.
- For cases above the mean on height *but* below the mean on weight: We have a positive times a negative number; this produces a negative number that increases in size the shorter or heavier the case is.

Now, if we summed these 'product of residuals' together across all the cases, we'd end up with a number that got progressively larger, to the extent that taller people tended also to be heavier. Our number would summarise how many cases we had in each corner or quadrant of our scatter plot. If we had a lot of cases in the bottom left and top right, and few in the top left and bottom right, we'd have a large number. If our cases were scattered randomly about the plot, we'd have a very small number. The formula for summing the 'product of residuals' is:

$$\Sigma(y-\bar{y})*(x-\bar{x}) \quad [\Sigma = \text{'sigma'} = \text{the sign for 'take the sum of'}]$$

This value is termed the *covariance* of our two variables. We have one final step to take. The value of the number we've just calculated will depend upon the units that each of our variables are measured in and the range of values for them in our data, and also by the number of cases we have. This is where the standard deviation comes in very handy. We multiply together the standard deviations of each of our two variables and then multiply the result by the number of cases (actually we take the number of cases minus 1, but we can safely ignore why we do so for now). If we divide the sum of our residual products we produced earlier by this number, it has the effect of 'standardising' our result for any range of values of any variable whatsoever. The final formula we have for measuring the association in our scatter plot is:

$$r_{xy} = \frac{\Sigma(y-\bar{y})*(x-\bar{x})}{(n-1)*s_x*s_y}$$

where

    $(n-1)$ is the number of cases minus one,

    s_x is the standard deviation of the variable x,

    s_y is the standard deviation of the variable y and

    r_{xy} is read as 'r for the variables x and y'.

The term 'r' is the *Pearson product moment correlation coefficient* named after its inventor Karl Pearson. We will encounter it again when we look at the analysis of variance and regression.

Pearson's r has truly remarkable properties. No matter what data is thrown at it, no matter what pattern exists in the data, it will always return a value between −1 and +1 for the association between two variables. Its absolute magnitude describes the strength of linear association, where 1 means perfect association and 0 means the absence of any association or 'independence'. Where values of both variables move in the same direction, r is positive. Where higher values of one variable tend to go with lower values of the other, and vice versa, r is negative. For our example of the height and weight of 11-year-old girls, the value of r is +0.66.

There is only one 'catch', but it is a very important one: r describes *linear* association only. That just means that the association does not change according to the values taken by either variable. As we'll see later, when that is the case, we can visualise the association as a straight *line*. Not all associations are linear however, but we'll deal with that complication later.

Conditional probabilities with a continuous and categorical distribution

The NCDS children sat a maths test at age 16. Figures 2.12 and 2.13 show the distribution of scores for boys and girls. Was performance in the test conditional upon their sex?

Comparing the two conditional distributions, it looks as if performance was conditional upon sex. The mean score achieved by boys was higher than that of girls, although not by a very substantial amount. This difference definitely exists, with a probability of 1. But there are two possible reasons for its existence. One is that there was an underlying difference in real maths ability as measured by the test between boys and girls. Another is that random variation accounts for the difference. A different sample of boys and girls of that age might have yielded different results. The *same* sample of girls and boys, were they to be tested again, would be unlikely to score exactly the same. Some children might be feeling unwell, distracted or just

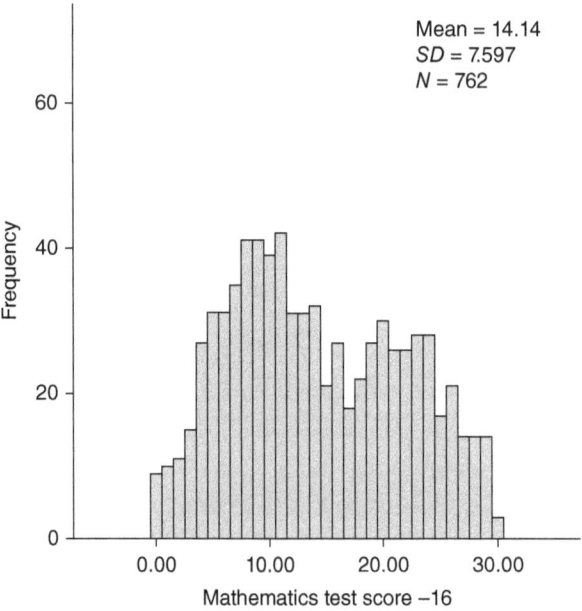

Figure 2.12   NCDS boys: maths ability at age 16

Note. NCDS = National Child Development Study. Data from NCDS teaching data set.

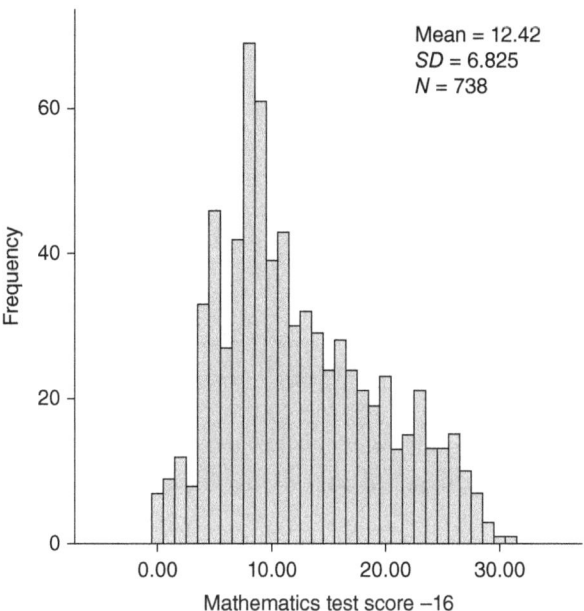

Figure 2.13 NCDS girls: maths ability at age 16

Note. NCDS = National Child Development Study. Data from NCDS teaching data set.

having an off day. Some almost certainly guessed the answers to some questions and got lucky – or *un*lucky! Thus, while we know for certain that there is a difference in the probability distribution in our sample, before we can generalise this to a wider population we need to consider the impact of such random variation. We look at this in the next chapter.

What have we done?

We have established a way of calibrating probability on a scale from 0 to 1. We have seen how to estimate probability as the *proportion of outcomes* towards which large numbers of identical independent repeated trials tend. We have seen how the resulting probability distribution can be described in terms of a table of frequencies, summary statistics (mean and standard deviation), bar charts or histograms. We have seen how we can compare theoretical or *expected* distributions of probabilities (e.g. the probability of *heads* on a coin flip, or a Gaussian curve with a mean and standard deviation) to the empirical distribution of probabilities that we actually observe. Along the way, we've established the following rather useful rules of probability:

1   The marginal probabilities of any outcome A in a sample space can only take a value from $0 \dfrac{\text{Zero Outcome } A}{N \text{ trials}}$ up to $1 \dfrac{N \text{ Outcome } A}{N \text{ trials}}$

2   The marginal probability of either outcome A *or* outcome B in a sample space will be the sum of their individual marginal probabilities.

3   The sum of all the marginal probabilities in the sample space of a trial must be exactly 1.

4   The marginal probability of outcome A in the sample space will equal 1 minus the combined probabilities or all the other outcomes. This is also equal to the probability of outcome A not happening.

5   The probability of two *independent* outcomes in two sample spaces is the product of their individual marginal probabilities.

Our next step in understanding inference is to examine *sampling distributions* and the construction of null hypotheses. To do that, we'll also look at the *binomial formula*.

Appendix: simulating coin flips with excel

You can use Microsoft Excel to simulate multiple Bernoulli trials. Excel has a random number generator controlled by the function RANDBETWEEN. In a blank sheet, select the top left hand cell (a1). Then, type *=RANDBETWEEN(0,1)* and press *Return* or *Enter*. You will see that Excel randomly places an integer between the range you have specified (0 and 1) in the cell. Drag the bottom right-hand corner of the green cell border to expand your cell selection downwards until you have selected a column of 100 cells. Excel will populate the 100 cells you have selected with randomly chosen 0s and 1s. In the cell immediately below your column of 100 cells, enter *=SUM(a1:a100)*. This gives you the total number of 1s in the random selection. Now, select all 101 cells and drag rightward for about 30 columns. This produces more columns of randomly generated 0s and 1s. You can use the other functions of excel to produce graphs or other summaries such as probability distributions of the series of trials you generate in this way. You can also simulate other trials, such as rolls of dice or any other trial set up by changing the terms of the RANDBETWEEN function.

To simulate a normal curve, use the *=SUM* function to produce a total for the randomly produced 1s (or any other range of numbers) across the rows or down the columns of a matrix of randomly produced numbers. You can then describe the resulting distribution as a frequency table, bar chart or histogram that will reveal an approximately normal distribution. You can treat each column in the matrix as a variable and the rows as values for those variables for each case observed.

When I did this, simulating about 30 variables, taking the value 0 or 1 for each observation, I got the results shown in Table 2A.1 and Figure 2A.1. The results are approximately normal. However, had I increased the number of cases, or the number of variables, the distribution would have become gradually smoother.

Table 2A.1 Excel simulation of a normal distribution

Score (= sum of 1s)	9	10	11	12	13	14	15	16	17	18	19	20	21
N cases with that score	1	6	1	7	14	14	18	10	15	3	4	4	2

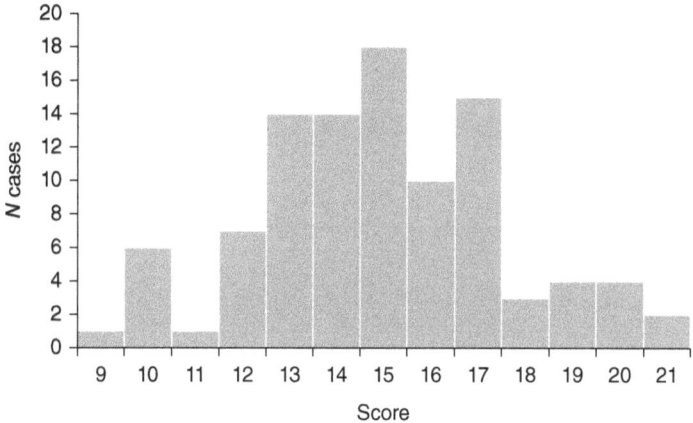

Figure 2A.1 Excel simulation of a normal distribution

Chapter Summary

- The probability of any event varies between 0 (impossible) and 1 (certain).
- We can describe the world in terms of variables that take values observed for individual cases in a data matrix.
- Variables can be described by the probability distribution of their values, which can be summarised by its level (e.g. the arithmetic mean) and spread (e.g. their variance or standard deviation).
- Trials take one or more mutually exclusive outcomes, which are comprehensively described by their sample space.
- Bernoulli trials take only two outcomes.
- The probability distribution of a variable conditional upon the distribution of another variable describes their association.
- The linear association of two continuous variables described by the Pearson linear correlation coefficient r takes values from −1 to +1.
- The Gaussian, normal or bell-shaped probability distribution is one in which most observations are found near the mean and become less frequent as the distance from the mean increases.

Further Reading

Michael Blastland and David Spiegelhalter's (2013) *The Norm Chronicles*, is an entertaining and accessible introduction to probability. As Winton Professor or Risk at the University of Cambridge, Spiegelhalter runs an excellent website about all things probabilistic (UnderstandingUncertainty.org). Ian Hacking's *The Taming of Chance* (1990, Cambridge University Press) sets the development of the understanding of probability in its historical context.

3

BERNOULLI, COKE AND PEPSI

Chapter Overview

A series of 10 trials

A friend remarks that they can always tell the difference between Coke and Pepsi. You do not believe her, so you conduct an experiment. You set out 10 glasses filled with one or other of the two drinks, after flipping a coin to randomly allocate which cola goes into each glass. You check that there is no clear visual difference between the two that would make their identification easy. Telling your friend that each glass might contain either drink, you invite her to identify the contents of each one, noting down whether she does so correctly or not (but not revealing the result until all the glasses have been tasted). She must reach a verdict on each glass. She may not say 'don't know'. If we insist on this, we can measure her skill quite independently of her own subjective views about it. At the end of the experiment, it turns out that she identified eight out of the 10 glasses correctly. Is this evidence that she can indeed tell the difference?

This might appear to be a frivolous problem, although, as we shall see, it has a rather famous pedigree. Working through the answer to this question allows us to establish all the essential logic of statistical inference. If you can follow this experiment and how to interpret its results, not only will you understand statistical inference, but you will also avoid the many traps that await anyone who tries to master inference by rote learning formulas or rules.

Statistical inference and research design

Before we proceed further, notice the nature of our test, and of the inference we hope to draw from it. When our friend made the claim, we might have 'tested' it in other ways. We could have asked her how it is that she could tell the difference, to describe it in more detail, or to account for how she came to acquire this skill. Such an approach might lead us to suspect a plainly false claim if the answers were evasive or lacking in detail. Conversely, no matter how eloquent our friend was, and no matter how extensive the account of her skill she might give, we would still feel obliged to carry out our test in order to follow the logic of 'Nothing on another's word'!

Notice also how little our test actually tells us, even if we interpret its results well. It tells us whether or not on this occasion, under these test conditions, our subject displayed an ability to discriminate between the colas we gave them. It does not allow us to infer what would happen under different conditions, although we might have many plausible ideas about what might be relevant. Most importantly, it tells us absolutely nothing about *how* any skill that we may find evidence for operates, where it came from, whether it is something anyone might possess, whether it is a function of some wider skill in discriminating all kinds of tastes and flavours or is it only about differentiating the two products we used. What happens *in* the tasting process that ultimately generates the data we analyse remains a black box, invisible to us. Our test only gives us data about its *results*. Our research design has had the aim of isolating the procedure of tasting from other factors. Our subject has no way of knowing which cola is in any of the glasses, and this is as true for the last glass as it is for the first. Nor does it depend upon the decisions they make on each glass. It could be that all the glasses contain the same cola, or that there is any permutation of the different colas in successive glasses. Under such conditions, we can distinguish any skill our subject has from how they might perform were they simply to guess their answers.

Finally, notice that this is nevertheless a potentially powerful tool. Were we testing the impact of a drug on patients, or consequences of some social experiment, rather than the taste of colas, improvement in the health or life chances of millions of people might rest upon the results.

The null hypothesis

Our drinks test is a series of 10 repeated identical trials, each of which has *only two* outcomes: Bernoulli trials. To answer our question of what a score of eight out of 10 in our series of trials might mean, we approach our problem in a rather roundabout way by asking what results we could expect from someone *who had no ability to distinguish* the two colas whatsoever. Their answers to the test would be random, in the sense that they might as well flip a coin to decide how to answer: if they had no ability to distinguish colas and the colas are served to them at random, there is no possible way they could improve on a score of guessing correctly half the time in the long run. Any answer they give has a 50% chance of being correct and 50% chance of being wrong. Each result is equally probable for each trial (correct guess, incorrect guess) so that we could expect exactly half their guesses to be correct in the long run. We have already established this from the law of large numbers that we saw in Chapter 2. This is the starting point from which we can formulate a **null hypothesis** about what would happen in our trials if someone with no ability to discriminate took them.

We know what would happen 'in the long run' with the null hypothesis – exactly half the guesses would be correct. But as Keynes quipped 'in the long run we're all dead', so what could happen in the 'short run', not of very large numbers, but small ones? In Chapter 2, we already had a glimpse of this process in Figure 2.5 which showed the very erratic estimates of the probability of heads that could come from a few trials, depending upon the early sequences of coin flips. We saw it also in the final tally of heads and tails after 10,000 flips. Each of our series of 10,000 flips ended up in a slightly different total number of heads. Now imagine what the range of possible totals for heads and tails might there be in such a series. It must range from 0 heads up to 10,000 heads. How many different possible sequences are there of heads and tails? There must be $2^{10,000}$. That is a number so large that, in effect, it is not a number. There is only one sequence of flips that leads to all heads, and one that leads to no heads. Each of these is just as possible as any other single sequence, but each individual sequence has the vanishingly small probability of $2^{-10,0000}$ of occurring: hence the gloriously erratic path of our running totals of the difference between observed and expected heads in Figure 2.6. However, there will be many billions of times more sequences that end up with roughly equal numbers of heads and tails than any other sequences. This is not due to any 'balancing' or evening up (remember, coins have no memory) but because there is a perfectly calculable distribution of the proportion of heads across all sequences. It comes from the **binomial formula** set out by Bernoulli. Formulas are for forgetting, but working them out from first principles is not hard, and it takes us right into the belly of the statistical inference beast and illustrates exactly how it operates.

Short run probabilities: the binomial distribution

What we often really want to know is, given some long run probability for an individual trial result, what results might I expect in the short run of a series of trials? This turns out to be extremely useful knowledge, because as well as moving between the short and long runs, we can use it to move between samples and populations. I can calculate the results *expected* in a short run of a small number of trials *given* any assumption I might make about the long run probability. I can then compare the results I observe with what I might expect to see, and from that, I can make inferences about what was going on.

First, we can work out how many different, equally probable sequences of trial results there could be from our 10 repeated cola trials. Since each individual trial has two possible outcomes, this must be $2^{10} = 1024$ (2 multiplied by itself 10 times). You can visualise this as a *tree diagram* like Figure 3.1, showing what might happen on the first five trials. Marked in bold is the start of one possible sequence with a correct identification (= **T**rue) of the first, fourth and fifth glasses, and incorrect guess on the second and third (= **F**alse). Each successive glass tasted doubles the number of possible sequences. By the time we get to glass number 10, there are 1024 branches to this tree, corresponding to all 1024 possible sequences.

Glass 1	F						T									$2^1 = 2$
Glass 2	F		T			F			T							$2^2 = 4$
Glass 3	F	T	F		T	F		T	F		T					$2^3 = 8$
Glass 4	F	T	F	T	F	T	F	T	F	T	F	T	F	T		$2^4 = 16$
Glass 5	F T F T F T F T F T F T F T F T F T F T F T F T F T F T F T F T															$2^5 = 32$

Figure 3.1 Tree diagram for trials 1 to 5

However, each of these equally possible sequences contains *different* numbers of correct and incorrect guesses. We can calculate how many of our 1024 sequences contains each possible total of correct guesses. There must be only *one* possible sequence with 10 *correct* guesses and *one* with 10 *incorrect* ones, because the only way to get all the trials correct is for each successive trial to be correct. The same goes for getting them all wrong.

TTTTTTTTTT

FFFFFFFFFF

There must be 10 sequences with only *one correct* result and also 10 ways of getting *one incorrect* one since the single (in)correct identification might come on the first or second or third … or tenth glass.

*F*TTTTTTTTT, T*F*TTTTTTTT, TT*F*TTTTTTT …

*T*FFFFFFFFF, F*T*FFFFFFFF, FF*T*FFFFFFF …

How many ways would there be of achieving two *true* results? Let us start with obtaining a true result on the first tasting. There would be nine tastings left in which the other *true* identification might occur. If our first *true* result came on the second tasting, there would be eight tastings left, on the third, seven tastings left and so on. Thus, the number of sequences with *two true* results must be as follows:

9 + 8 + 7 + 6 + 5 + 4 + 3 + 2 + 1 = 45 sequences

For *three true* results, we will need to go through the same procedure as for *two true* results, but this time for the number of tastings left after the first two true results have taken place.

Clearly, this is a tedious way of proceeding, but also one that we can simplify with the formula Bernoulli discovered, since we are repeating essentially the same procedure. This involves many calculations of the form 4 * 3 * 2 * 1 or 6 * 5 * 4 * 3 * 2 * 1.

Maths simplifies the presentation of these calculations by referring to them as *factorials* and writing them as 4! or 6!:

$$4! = 4 * 3 * 2 * 1$$

$$7! = 7 * 6 * 5 * 4 * 3 * 2 * 1$$

The **binomial coefficient** gives us the number of ways of arranging k results within n repeated trials as follows:

$$\binom{n}{k} = \frac{n!}{k!(n-k)!}$$

You read $\binom{n}{k}$ as 'n choose k'. Test this out for two correct results:

$$\binom{10}{2} = \frac{10!}{2!(10-2)!} = \frac{10*9*8*7*6*5*4*3*2*1}{2*1*8*7*6*5*4*3*2*1}$$

$$= \frac{10*9*8*7*6*5*4*3*2*1}{2*1*8*7*6*5*4*3*2*1} = \frac{90}{2} = 45$$

Factorials are easy to cancel out when they appear on opposite sides of a fraction. *Binomial coefficients* are extremely useful in any situation when you have to figure out how many ways there are of doing different things, or of the number of combinations of choices implied by a series of options, menus or decisions. The formula is always the same and takes the general form:

$$\frac{\text{(Total number of places)!}}{\text{(Number of places to be filled)!} * \text{(Number of places left)!}}$$

Note that the *sum* of the two numbers in the denominator equals the number in the numerator. In our example, 8 and 2 is equal to 10. We can now use the binomial coefficient to work out the number of possible patterns of trial results with each total number of correct cola identifications from 0 to 10. If we do so, we get the results given in Table 3.1.

Table 3.1 Distribution of total correct guesses

Number of Correct Guesses	Number of Equally Probable Trial Sequences
0	1
1	10
2	45
3	120
4	210

Number of Correct Guesses	Number of Equally Probable Trial Sequences
5	252
6	210
7	120
8	45
9	10
10	1
All	1024

If you study Table 3.1, you will see that we have created a **probability distribution** for all the 11 possible results of a series of 10 trials (zero through to 10 successes). All we need to do to convert these frequencies into probabilities is divide by the total number of sequences (1024) to get a distribution of probabilities. These will sum to exactly 1 because they describe the entire sample space.

Bernoulli also set out the *binomial formula* which translates the frequencies from the binomial coefficients directly into probabilities. In a *series* of length n of repeated identical *Bernoulli trials* in which the (law of large numbers) long run probability of an outcome in any single trial is P, the probability that an outcome will occur k times in n trials is given by

$$\frac{n!}{k!(n-k)!} * p^k * (1-p)^{n-k}$$

This formula appears rather daunting, especially if you are not used to dealing with factorials or powers, and there is no need whatsoever to remember it (*in real life, a formula is never further away than the nearest web browser*). We can use the formula to answer our original question about the probability of obtaining different numbers of *successes* in a series of 10 trials, where the long run probability of getting a correct identification by guessing must be 0.5. Let's check what we obtained before for a score of 2 out of 10, which was 45/1024 = 0.044

$$n = 10, k = 2, p = 0.5$$

$$\frac{n!}{k!(n-k)!} * p^k * (1-p)^{n-k}$$

$$= \frac{10!}{2!(10-2)!} * 0.5^2 * (1-0.5)^{10-2}$$

$$= 45 * 0.5^2 * 0.5^8 = 45 * 0.5^{10} = 0.044$$

If you examine the workings of the formula, you will see that it takes the binomial coefficient that, as we saw earlier, provides the number of sequences that yield each total of true results, and then divides it by the total number of possible sequences. Thus, we are back to our general definition of probability again: the frequency of an outcome across a number of trials. Note that this formula is good for *any* long run probability, not just for trials where outcomes are equally possible. We'll return to that later.

Using the binomial formula, we get the results shown in Table 3.2. You might want to check if you get them too. There is no need to do the calculations for 6 through 10, since these must be the mirror of the results from 0 through 5. If I have four true results, that must mean six false ones, and so on.

Table 3.2 Probability distribution for correct guesses if H_0 is true

Number of True Results	0	1	2	3	4	5	6	7	8	9	10
p		0.001	0.010	0.044	0.117	0.205	0.246	0.205	0.117	0.044	0.010	0.001

You can also get the same results using 'Pascal's triangle', reproduced in Figure 3.2.

```
              1
            1   1
          1   2   1
        1   3   3   1
      1   4   6   4   1
    1   5  10  10   5   1
```

Figure 3.2 Pascal's triangle

In the triangle, each number is the sum of the two numbers above it (if it is at the 'edge' of the triangle, there is only one number above it, and it is always the digit 1, so that we just add this to zero). If you study the triangle, you will see that each number also represents the number of routes through the triangle that can be taken to arrive at that point. We can use these numbers to represent the number of sequences in a series, where the length of the series corresponds to the rank in the triangle below the apex. The location of the point along the horizontal axis of each layer of the triangle, counting from the left and starting at zero, yields the number of outcomes of inter- est. Thus, the second rank of our triangle shows that there is one sequence that yields zero successes and one sequence that yields one success. The third rank shows that there is one sequence that yields zero successes, two sequences that give one success and one sequence that gives two. The final level shows that there is still only one sequence giving zero successes from five trials, five sequences giving one, 10 giving two, 10 giving three, five giving four and only one giving five. Pascal's triangle is sim- ply a visual analogue of the binomial formula. The 19th-century polymath Francis Galton constructed a physical version called a 'quincunx' using pins precisely nailed on a board through which lead shot could be dropped, which bounced randomly to the left or right of each pin. There are many virtual quincunxes available on the internet, such as at www.mathsisfun.com/data/quincunx.html.

The sampling distribution and *p*-values

It is worth taking some time to consider what Table 3.2 shows. It is special case of a *binomial probability distribution* called a **sampling distribution**. It shows the probabilities of getting each of the 11 possible outcomes from *one* series of length *10* trials, if the *long run* probability of success in each individual trial is 0.5. It is therefore the sampling distribution of *k* successes for a series of length $n = 10$ trials with $p = 0.5$. Take your time to pay close attention to the following eight features.

1   *The sampling distribution describes the results expected under the null.* It describes the sample space of the results of 10 trials and their probability distribution under conditions of 'no skill' or the *null hypothesis* or H_0. We have produced an account of what our data from the experiments would look like *if the null hypothesis was true*. To use the language and notation of conditional probability that we saw earlier, our histogram is a description of

$$P(\text{data} \mid H_0 = \text{true})$$

2   It is a *sampling distribution* because we can treat each of the individual, equally probable, sequences of 10 trial results as a single example available to be randomly selected from the population of all 1024 sequences. It gives us the frequency of successes and associated probabilities we would *expect* to see, *if the null hypothesis is true*. We can see that some of these sequences are very unlikely to occur if the null is true, and the associated probabilities are very small.

3   *The probabilities sum to 1*, as they must do, because they describe *all* the mutually exclusive possible results from a series of 10 trials, and the probabilities of getting them were we able to repeat this series of 10 trials again and again. Any individual series of 10 trials that we might run is *one* sample from *all* the equally probable possible series of trials of that length that might ever exist. In exactly the same way as we would be surprised to flip a coin 10 times and obtain nine or 10 heads, but think nothing of getting four or five, we'd normally expect clueless cola tasters to get roughly half the trials right, not most of them. They would be very lucky indeed to identify nine or 10 glasses correctly. The probability would be about 11/1024 = 1.1%. From the law of large numbers, we could say that in an imaginary world where we repeated this 10-trial experiment, many thousands of times with subjects who had no ability to discriminate between the two drinks, we could expect the long run proportion of times that they scored either nine or 10 correct identifications to approach 11/1024. By the same logic, we'd expect them to get *at least* eight out of 10 correct on 4.4 + 1.0 + 0.1 = 5.5% of attempts.

4   *We refer to these probabilities as **p-values**.* They describe the probability, assuming the null hypothesis is true, of observing a result, or one more extreme than that, such as 'at least nine correct identifications'.

5   *The probabilities are approximately normally distributed.* Figure 3.3 represents the distribution as a histogram. Does the shape of our histogram seem familiar? With only 11 values on the horizontal axis, the approximation is not very close, but you can see that the mean and median number of successes coincide at 5, and the probability of expected results declines as we move above or below that number. Like any distribution, this one has a standard deviation as well as a mean. The formula for the standard deviation of a binomial distribution is as follows:

$$SD = \sqrt{N * p * (1 - p)}$$
$$= \sqrt{10 * 0.5 * 0.5} = \sqrt{2.5} = 1.58$$

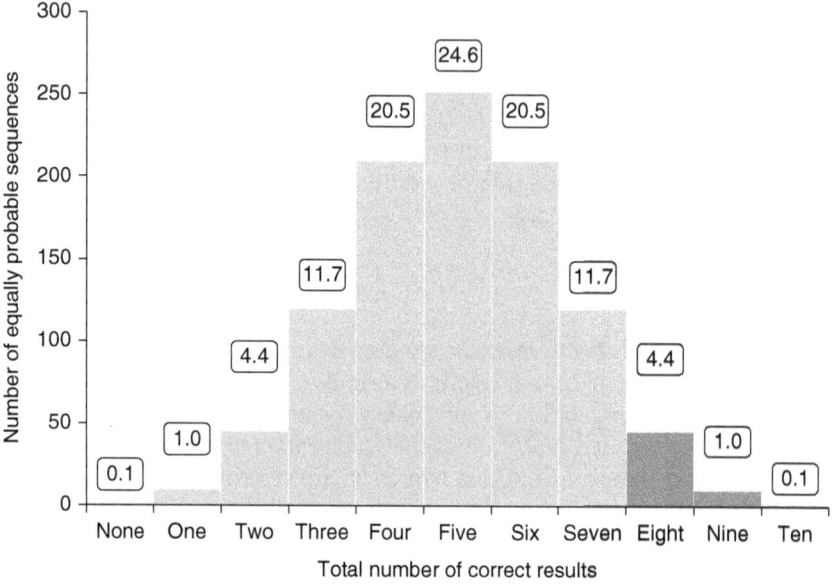

Figure 3.3   Probability distribution for results $n = 10$, $p = 0.5$

6   *The standard deviation of the sampling distribution is called a standard error.* Like any standard deviation, it is a guide to how closely values cluster towards the mean, or disperse far from it.

7   *This sampling distribution is an imaginary one that only ever exists in our heads.* In reality, the mere fact of repeating the experiment so much might be that our subjects learn to discriminate between the drinks or become so sick of them that they are unable to face them. That would mean that our trials would gradually stop being identical. We used randomisation to allocate the colas to each glass in our real experiments in order to replicate our imaginary random world as closely as practically possible. We only need to carry out *one* series of trials to compare it with the infinite number that we have conjured up in our minds.

8   *The distribution contains one or more* **critical regions** in which we may decide the probability of the results is so low under the null that we have enough evidence to *reject* the null hypothesis. The portion shaded black in Figure 3.3 describes this region, if we decide that a *p*-value of 0.055, corresponding to identifying eight out of 10 glasses correctly, is sufficiently low to lead us to reject the null.

Alpha, significance, confidence and type I errors

This sampling distribution can be used in an ingenious way if we reason as follows. We cannot know what the sampling distribution of the trial series results would look like for people who *can* tell the difference between colas, because we don't know anything about the skill or its distribution in the population. Maybe even the most expert, highly skilled cola tasters only manage to identify the right drink six or seven times out of 10. Or perhaps, even rank amateurs have no trouble telling the difference almost every time. Maybe almost everyone has the skill, or almost no one, or maybe there is a range of cola discrimination abilities in the population. However, we *do* know what the results *would* look like for people with *no* ability. This is what we have just worked out with our sampling distribution of test results under the null! Therefore, we could decide we had *provisional* evidence of *the existence of ability* for subjects who achieved a success rate that would have a very *low probability* of occurring for those with *no* ability.

Take some time to think through this double negative: *not* probable with *no* skill is provisional evidence of some skill! Thus, if we defined identifying eight or more out of 10 glasses correctly as provisional evidence of skill, we could do so knowing that each time we ran the test, there would be only a 5.5% chance of someone with no skill at all meeting this standard by luck. We could therefore feel fairly confident in defining anyone who managed this feat as being capable of cola identification. The 5.5% of tests of clueless cola drinkers in which they scored 8 or more out of 10 by luck would be **false positives**. They describe the probability of observing each possible result, or one more extreme, *conditional upon the null hypothesis being true*. What we have done is to take 5.5% or 0.055 as an acceptable risk level for falsely describing a lucky but clueless taster as a cola identification expert. This is also known as a **Type I error**: *rejecting* a null hypothesis when it is in fact true. The Type I error rate that we're prepared to accept is known as the **significance** or **alpha** (α) level. In some experimental or survey designs, it makes sense to define this level *before* examining the data, so that we are not tempted to nudge it up or down when we see our results. We defined what we would *expect* to see if the null is true. If we *observe* something sufficiently different to this, that is, something sufficiently *im*probable under the null, we have evidence to *reject* it. We can also use our *p*-value as a description of how confident we are about the inevitably risky inference we have just reached. Because we think it is a procedure that would throw up a false positive 5.5% of the time, we could say that we were 100 − 5.5 = 94.5% *confident* about our result.

Conversely, if we decide that eight out of 10 is not convincing enough, we could declare a negative result, and *accept* the null. We might decide to place the rejection level at 1.1% and insist on nine correct identifications. Here too, we face two possibilities. If in fact our friend is clueless but lucky enough to score eight, we have a *true negative*. In such a scenario, the null may be true, and we happen to be correct to accept it, but we haven't obtained very strong evidence about it. Have we proved that the null hypothesis is true? That is *not* a conclusion we can safely reach. If on the other hand, our friend really could tell the difference (she just made two slips), we have a **false negative** and have committed (you've guessed it) a **Type II error**. Perhaps, even fantastic tasting skills do not improve the ability to identify the colas so dramatically as to regularly get nine out of 10 correct. Notice that whatever we do, given the size of our sample, to drive down the risk of a false positive result, will increase the risk of a false negative one, and vice versa.

Improbable expectations, definite results

At this point, keeping a clear head, and remembering Union soldiers and Feynman's number plate is useful. There is nothing probable about the result we obtained when our friend correctly identified eight colas. This result, our data, is as real as Feynman's number plate. Its probability of existing is one. Nor is there anything probable about our friend's actual ability to discriminate between colas. Again, with probability of 1, *either* she can do this *or* she cannot. She cannot 94.5% tell the difference, nor is there 94.5% of her that can and so on. Both her and her skill are indivisible. *What is probable is the state of our knowledge about her skill.* Here we have conflicting evidence. She correctly identified eight out of 10 tests, which looks pretty good. However, we also know that someone *without* any skill at all would get this score on about one out of 20 occasions in the long run, given our testing set-up. Such a skill-less person is equivalent to a very short or very tall Union soldier. Just as most Union soldiers were close to average height, most times a skill-less person tasted the colas, they would get around half right, but every so often, just as we might pick a very tall Union soldier, they might get a score of 8 or 9. The probability that we have calculated does not tell us how likely it is that something is true, given the result we obtained. It tells us the probability, in the long run, of getting such a result *when the null is true*. This probability does not refer to our single result, it refers to the probability of getting such results *in the long run*. It tells us about our *procedure*.

Why the double negative and all the bother?

The confusion arises because normally we are interested in a substantive result, such as 'Person X *can* tell the difference between Coke and Pepsi' or $P(H_{cola} \mid data)$. Our theories might be 'Smoking causes lung cancer', 'Minority ethnic applicants are discriminated against in recruitment' or 'Working-class men vote Republican' and we want to know the probability of them being true. We want to be able to say something like 'it is 95% probable that my friend can tell the difference between Coke and Pepsi', rather than what we can in fact say which is 'on only around about 5.5% of occasions, using these procedures, will someone without the ability to discriminate be classified wrongly as having that skill'. Why this unwieldy and confusing double negative? We are unable to go further because we lack two important pieces of information. The first is the prevalence of cola identification skills in the population. Maybe almost everyone can do it. Maybe hardly anyone can. Our second problem is that we do not know the size of the effect we are trying to measure. Maybe it is large and sends a strong signal: anyone who can tell the difference can do so every time with little difficulty. But maybe it is weak. Perhaps even the most expert drinkers can identify the right cola only 60% or 70% of the time. If we *do* have access to these two pieces of information, then we can go further. We examine that scenario in Chapter 6 when we look at power.

The logic of NHST

This chapter has taken you through the logic at the heart of statistical inference: NHST. The remaining chapters in this book only elaborate on the basic ideas here, apply them to different contexts or derive other insights from the same underlying logic. Master this logic and you will understand statistical inference. The logic is simple, but it is also counter-intuitive and it's easy to slip up. This helps explain why it took about 250 years to go from the discovery of the binomial formula by Bernoulli to the logic of inference and random sampling developed by F. Y. Edgeworth, Karl Pearson, R. A. Fisher, Jerzy Neyman and others in the early 20th century. Let's review what we have achieved.

The first, and relatively trivial, insight is a more accurate view of the probabilities of success on a series of trials compared to an intuitive guess. 'Getting it right at least half the time' over 10 trials sounds quite good until you realise that with no skill one could expect to achieve this about two thirds of the time! (if you are not sure why, review Figure 3.3).

The second achievement was to successfully measure something *that was not only invisible, but might even not exist!* In fact, the ability to discriminate Pepsi from Coke is not something we could easily measure *in any other way*. We could ask people about their skill, and they might try to tell us, but how could we (or they) know if they were telling the truth? They might genuinely believe that they possessed this skill but be unable to 'prove' it. Instead, we set up an experiment within which only real possession of skill was likely to produce a pattern of results. We used randomisation in two ways. We used it to allocate trials to subjects, so that they could not know in advance whether it was more likely that any individual trial would have one or other result. We also used it to calculate the probability distribution for results given a true null hypothesis. We then used this knowledge to decide when a pattern of results was sufficiently unlikely to be produced by randomness alone as to constitute evidence of some other process at work, such as skill at differentiating between colas. We were able to do this *without any prior knowledge* of what results Coke discrimination skill might produce.

We would be on dangerous ground if we took the results of one set of trials as robust evidence of the existence of this skill. However, if we found that we could replicate these or similar results with other subjects, we could more confidently declare the existence of this invisible process.

Our third achievement was that we required *no previous knowledge* about cola tasting skill in order to carry out our experiment. We did not need to know whether it existed or not, how strongly developed any such skill might be in a person possessing it or what its prevalence might be in the population. We did, however, have to have enough of an empirical sense of how it might manifest itself to set up our experimental conditions. We had our subject *drink* glasses of cola, we didn't have her look at, listen to or weigh glasses of cola or perform some other test.

Our fourth achievement was to measure a target *population*, based on a *sample* taken from it: we *generalised* from the sample to the population. We gave our subject only 10 repeated trials. These 10 trials were a sample from the infinite super population of trials she could ever have taken. The sampling distribution we produced for a series of length 10 trials depended on only two things: the feature we were trying to measure and the number of trials in the series, or what amounts to the same thing: the size of the sample. This is revealed in the formula we saw for the standard error of the sampling distribution.

Note three things about it:

1   Error doesn't mean *mistake*. It just describes how samples tend to vary.
2   This variation *is not related at all* to the size of the population.
3   It depends on the *square root* of the sample size. This means that, for example, halving the size of the standard error requires a sample four times larger.

Basically, the size of the population doesn't matter. It can be thousands, millions or trillions. The accuracy of estimates we make from our sample only depends on the size of the sample itself and not on the ratio of the size of the sample to the population.

From statistical inference to scientific inference

Our calculations have dealt with the statistical modelling aspect of inference. But we are not quite done with the whole process of inference itself. So far, we have dealt with an imaginary world in which people not only happily taste countless glasses of fizzy drink but also approach each glass with their taste buds as fresh as they were for the first glass. Not only that, to ensure an accurate trial, we would need to ensure that every single serving of cola was identical for every trial for every participant, save for whether it contained Coke or Pepsi. We'd also need to ensure consistency in the test circumstances. Maybe the ambient temperature, light or noise level affects the tasters, or the time of day, or season. In other words, in the real world there will be many sources of variation that we may be unable to control or even anticipate, all of which might have some impact on our results. Our purely statistical model only tells us about what happens when everything is optimal. In the real world, things are not that perfect, so that we always have to consider research design.

What does *significant* mean?

The statistician who did more than any other to establish the usefulness of NHST logic was R. A. Fisher. He applied his mathematical skills to the analysis of results of testing different treatments and growing conditions for crops at the government agricultural research station at Rothamsted in the early decades of the 20th century. He transformed what had been ramshackle trial and error into experimental procedures that came to be widely copied across every scientific discipline through his book *Statistical Methods for Research Workers*, published in 1925 and still in print today (Fisher, 1925). The challenge he faced was how to distinguish the vagaries of sunlight, rainfall, weather conditions, infestations by pests, soil quality and dozens of other variables that might affect crop performance from those factors researchers wished to study. The solution was randomisation, but this still required some way of determining whether any particular treatment had an effect promising enough to stand out from random variation. He used the term *significant* to describe any result in some data that had a sufficiently low probability of being observed, *if* the null hypothesis that a treatment had no effect was in fact true. He argued that it was best to decide at what level this probability should be set before carrying out the experiment and obtaining the results, but he also implied at times that a level around 5% was often a convenient compromise between catching too many false positives (rejecting a null hypothesis and declaring some result significant when the null was in fact true) and losing too many false negatives (accepting the null, when in fact there was a real effect at work).

However, Fisher was clear about five things. First, he used the term *significance* to mean 'signal'. A significant result was one that *signalled* to the investigator that something was probably happening (as it would be in our example if our friend did have skill in differentiating colas), and that the data was more than simply random noise (as it would be in our example if our friend had just made lucky guesses). The English language is constantly evolving. Today, the word *significant* has morphed into meaning 'important'. The word did not have this connotation in Fisher's day and Fisher did not use the term in that sense. Second, he stressed that one significant result meant little. It was a signal to the investigator that the result *might* be a fruitful area for further research. This might take the form of repeating the experiment to see if the result occurred again, or trying new ways of examining what might explain the data the researcher had. Third, he stressed that rejecting a null hypothesis did *not* mean accepting (let alone proving) any specific alternative hypothesis, although it might strongly imply that such a hypothesis was likely. In our example, where we divided the population into cola experts and others, deciding it was unlikely that an individual was an 'other' on the basis of rejecting the null hypothesis, does strongly imply that an alternative hypothesis of 'this person can differentiate Coke and Pepsi' *is* true. Fourth, he was equally clear that *not rejecting* a null was not at all the same as confirming or even accepting it. It could simply mean that the experiment had not produced good enough evidence either way. Finally, he was crystal clear about the meaning of the *p*-value. It is the *probability of obtaining the data observed*, given the data *we could expect to observe if the null hypothesis is true*. It is NOT the probability of the hypothesis being true, given the truth of the data, or given the data obtained.

Appendix: A famous experiment: milk first or tea first?

Some tea drinkers claim that pouring the tea first results in a poorer taste because the hot tea scalds the milk as it is added. Conversely, adding tea to the milk heats it more slowly as the first drops of tea are cooled by the milk. The story goes that one of the guests at Fisher's social gatherings had claimed to be able to tell the difference (versions of the story differ but the events do seem to have taken place). Fisher prepared eight cups of tea, four with the milk poured first, four with the milk added after the tea was poured. The preparation and presentation of the cups were randomised. The subject was told that four of each kind of cup of tea had been prepared and was invited to identify them.

We cannot use the binomial formula to calculate the probabilities in Fisher's experiment. The eight tests are no longer identical, because the subject knows the number of cups with each condition. For example, having chosen to identify three cups as being of a particular type, the subject would have only one cup of that type left to allocate, and so on. Were they to have no cups left to try after identifying three of a kind, they'd have to reconsider their earlier choices, as they would if they initially suspected five cups of fulfilling one of the conditions, and so on.

However, we can set up a contingency table (Table 3A.1) to describe the possible results of their choices (there are five) and calculate the number of equally possible permutations producing these choices were they driven by luck rather than by judgement, as with the cola test. There are 70 permutations.

Table 3A.1 Pairs of tea cups

		Real Condition		
		Tea First	**Milk First**	
Subject's Judgement	Tea first	0–4	0–4	4
	Milk first	0–4	0–4	4
		4	4	8

The first (top left) cell in the table can take five values from 0 to 4. Given the value in that cell, the values in the other cells are determined by the need to arrive at the marginal totals. Thus, if the subject correctly identified three 'tea first' cups, they must also correctly identify three 'milk first' ones, and incorrectly identify one each of a 'tea-first' and 'milk-first' cup.

It is useful to think of four pairs of cups, with a 'tea-first' and 'milk-first' cup in each pair, since the taster knows that for each cup of one type that they identify, there is also a cup of the other type, which we could think of as its 'twin'. Let's identify the cups with letters, with A to D being 'tea first' and E to H as 'milk first'. The probability distribution will be as follows:

1 permutation of choices that correctly identifies four pairs

1 permutation of choices that correctly identifies no pairs

Identifying A to D all correctly as tea first, logically implies that E to H must be milk first. Imagine tossing four coins: there is only one way to get either four heads or four tails. Conversely getting all A to D wrong must mean getting all of E to H wrong too.

There will be 16 ways to correctly identify only one pair. Each of A to D can be chosen correctly, and for each of those single correct choices, there will be four partners E to H to match that choice (AE, AF, AG, AH, BE, BF, . . .). Again, the number of ways to *in*correctly identify only one pair must be the same. So that we have the following:

16 permutations of choices to correctly identify one pair

16 permutations of choices to correctly identify three pairs

Finally, there will be six ways to correctly identify two from A to D correctly (set out below) and for each of these, six ways to match these choices from the milk first cups. Thus, there will be 36 permutations of choices to identify two pairs correctly:

ABCD   **A**BC**D**   **A**BC**D**   A**B**C**D**   A**BC**D   AB**CD**

We could represent this as a histogram. With four pairs of cups (Figure 3A.1), there would be a 1.4% (1/70) chance of identifying every cup correctly by sheer luck. In the event the guest did identify all the cups correctly and we could conclude that she did in fact have the capacity to tell the difference, with a risk of committing a Type I error of 1.4%. This probability expresses the fact that if we were to repeat this test many times with subjects who had no ability to tell the difference and were guessing, we'd expect about 14 such subjects to hit the jackpot of four correct guesses for every 1000 trials that we ran.

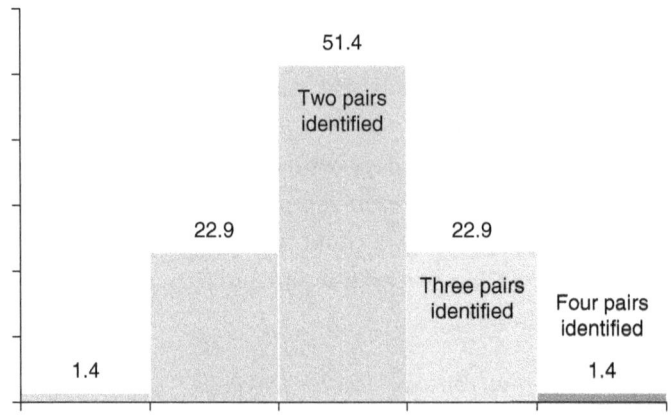

Figure 3A.1 Four pairs of cups, $p = 0.5$

Note that while the test has a low Type I error rate, the small number of trials means that a Type II error is again quite likely. Someone with no ability could correctly guess three pairs almost a quarter of the time, so that would not be a stringent enough criterion to accept evidence of ability. But demanding four correct guesses leaves no room at all for the subject to slip up. Someone with an almost perfect ability to identify whether the tea had been poured first would fail our test if they made only one mistake.

Activity: Key Terms

This chapter has covered a lot of ground and introduced a number of technical terms. Check that you understand them by trying to write a short description of each.

alpha

binomial coefficient

binomial formula

critical region

false positive

false negative

null hypothesis

probability distribution

p-value

sampling distribution

significance

Type I error

Type II error

Chapter Summary

- A null hypothesis about a target population is one for which we can calculate a sampling distribution for a sample statistic that estimates the value of a population parameter.
- The standard deviation of a sampling distribution is termed the *standard error*.
- The binomial formula describes the probability distribution of the results for a finite series of repeated identical independent Bernoulli trials.
- *p*-Values describe the probability of obtaining a range of results conditional upon the null hypothesis being true.
- For a given significance or alpha level, a critical region of the sampling distribution can be described where results imply rejection of the null hypothesis.
- The significance level or *p*-value for a result is also the long run probability of committing a type I error, or declaring a false positive.

Further Reading

There are innumerable texts describing NHST, some worse than others, but remarkably few that manage to combine rigour with accessibility. *Statistics* by David Freedman, Robert Pusani and Roger Purves (W. W. Norton several editions from 1978 onwards) is one of the best, as is *Statistics* by Alan Agresti and Chris Franklin (Pearson, 2007). David Salsburg *The Lady Tasting Tea: How Statistics Revolutionized Science in the Twentieth Century* (Owl Books, 2002) is a history of statistics told through brief biographies of leading statisticians. A brief but incisive account of Fisher's activities at Rothamsted is Simon Raper (2019) 'Turning Points: Fisher's Random Idea' in *Significance, 16*(1), 20–23.

4

SAMPLES AND POPULATIONS

Chapter Overview

With the concept of the sampling distribution of a statistic under the null, and the logic of NHST under our belt, we are now ready to think about samples and populations. First, let's remind ourselves about why we need samples and why they should be random. We use samples, as we saw in Chapter 1, because populations are too big to measure; because they may not yet exist, like future patients of a health treatment or citizens subject to a forthcoming change in the law or because they will only ever exist in our imagination, as part of a logical thought experiment. Usually we are interested in measuring the present or past, and that is certainly all we can possibly do, but when we wish to do any more than tell historical stories, we need to be able to generalise to the future. Investors are always warned that 'past performance is no guarantee of future results'. However, our purpose as social scientists is usually to make statements not just about what has happened in the past but how we therefore expect the structures or processes we investigate to behave in the future. This means that even in situations where we are measuring an entire *existing* population (e.g. patients in a drug trial, children in a school or long-term unemployed on a training course), we will want to treat it as a *sample* from the population of all the people who will be in that category in the future (people prescribed the drug, children attending the school next year or the future long-term unemployed).

The difficulty of measurement

When we think of 'measurement' we tend to think of a ruler or tape measure, or perhaps a thermometer. Getting a rough picture of the temperature where we are or whether a new piece of furniture will fit in a space is straightforward. However, few of the variables that social scientists are interested in are visible. Allocating someone to a social class may require a long inventory of information about their occupation (if they are employed) or work in the past if they are not, their current and past family and household status and income and ownership of assets. Valuation of the assets may pose further challenges. Placing their occupation in a hierarchy of skill or prestige will require knowledge about the (dynamically changing) occupational structure of the economy or labour market they are part of. Is that market local, or international? The UK Labour Force Survey collects data on the skills and economic activity of around 60,000 households each year, principally in order to provide evidence about employment trends. That might appear to require a few questions about whether respondents are working, their skills and education, occupation, hours of work, earnings, whether they receive state benefits of various kinds and the composition of the household. In fact, each interview generates up to 900 variables, and further variables are derived from these. There are 15 volumes of 'User Guides' to the survey and its methodology. Despite this, there are many questions about the UK labour force that the Labour Force Survey is unable to answer.

Because measurement is difficult, time-consuming and expensive to do well, it is almost always better to measure a small sample of what we wish to describe, and use it as a guide to the target population which is the real focus of our interest. But how do we know if that sample is any good or how big it should be?

The illusion of representative samples

Over the last few hundred years, many different schemes have been suggested for devising 'representative' samples. Often they depended upon using expert knowledge of some kind to construct a group of examples that were imagined to be typical in some way. None of these schemes worked. People often talk of 'representative' samples. This is a comforting idea. If we have a *sample* that properly 'represents' a *target population*, we should be fine measuring the sample and generalising our results to the population from which it has been drawn. But schemes for representative samples all fail to resolve what appears to be a logical impossibility. Just how good a representation of the target population any 'representative' sample might actually be could only ever be determined by comparing the sample with the population. But it was precisely the difficulty or impossibility of measuring the population in the first place that was the only reason to draw a sample! Even if some of the features of a target population were known (e.g. the proportion of men, or those aged under 20 years), this could be no guide to *other* features. It would be perfectly possible to draw a sample with the correct proportion of men or young people, which nevertheless contained a much larger or smaller proportion than the target population of any other features we wished to measure. And if we already knew about some features of a population, it would only be the information about those unknown features that we could possibly want the sample for! There would be no means of checking! Representativeness in this sense was, and is, despite the protestations of those who deal in 'ideal types', an illusion. Unfortunately, it is an illusion that continues to motivate the selection of informants in much qualitative work and in developing panels of 'online' respondents.

It has also led to some well-known survey disasters in the past. A famous example came in the 1936 US presidential election. As it had done for every election since 1920, the *Literary Digest* sent out some 10 million straw ballots on the presidential contest to names gleaned from automobile registration lists and telephone books. Some 2.3 million papers were returned, with 55% for Alf Landon, the Republican candidate, and only 41% for Roosevelt. In the real election, Roosevelt not only won but by the biggest victory in terms of electoral college votes in any US presidential election before or since, gaining no less than 61% of the vote, to Landon's 37%. The *Literary Digest* got it badly wrong because Democrat voters were less likely to have cars or telephone lines. Meanwhile, a young pollster, George Gallup, made a name for himself by correctly predicting the election result using a poll based on a much, much smaller but more genuinely random selection of voters and correcting for non-response.

Such problems recur in every 'readers' survey' or the online polls frequently run by papers and magazines. Those who are willing to make the effort to respond are overwhelmingly those with a keen interest in the issue at stake. Their views may be important, but there is no reason whatsoever to expect them to reflect those of the rest of the population. A good example of this is an oft-quoted statistic from a Shere Hite book '*Women in Love*' that 70% of women married for at least 5 years have had an extra-marital affair, 'based on a sample of 4500 women'. However, to obtain this sample, 100,000 questionnaires were issued. It's probable that women with a motivation to participate in the survey (perhaps *because* they had experienced an affair) were over-represented in the 4.5% who returned a questionnaire. Enter Atkins et al. (2001) who used US *General Social Survey* data (collected through a face-to-face interview on a *random* sample of households) to examine the same topic. Their estimate of the rate of marital infidelity by women after 5 years of marriage – 5%!

Marketing copy often uses samples such as '9 out of 10 doctors recommend ...', '9 out of 10 cats prefer ...', '99% of people said ...'. The only logical responses to such claims is 'Which doctors?' Nine million out of 10 million doctors, or the only 9 doctors on the planet who recommend something, plus 1 other to make the number more convincing? If you pick and choose who or what gets to be in your sample, you can get any result you like. The publicity and marketing industry is well aware of this. It is also well aware that any statistic, no matter how worthless, lends an air of scientific kudos to a claim. The moral of the story: if the advert, news article or other output does not provide a link to background information describing how the 'survey', experiment or test was carried out, and this information demonstrates clearly that the sample was a random one or a near substitute, then treat its 'findings' as nonsense. A point to ponder: how might one obtain a random sample of cats, dogs or any other household pet for that matter? If convenience samples or 'representative' samples are worthless, what does a random sample offer? To understand this, we need to think further about sampling distributions.

The sampling distribution, samples and populations

Let's return to the experiment we considered in Chapter 3. We created a *sampling distribution* of k, the number of successes in a series of 10 trials, assuming that the null hypothesis of no skill was true, and calculated the *p*-values associated with each value of *k* in that sampling distribution. We could call the different possible values of *k* and their associated probabilities a *sampling* distribution because

- it described the results of *every* imaginable possible series of 10 trials,
- it assigned a probability to each possible result and
- it could be calculated using pure logic, with no empirical evidence.

This last point needs some attention. Empirical knowledge was involved in the very limited sense that we needed to assure ourselves that the trials were independent of each other, were identical, involved only two possible outcomes and, in the absence of any skill, a random guess was the best strategy of the subject. However, we did not need to know anything about the empirical features of the trials, or how subjects might perform in them in order to create our sampling distribution.

But why do we call this distribution a *sampling* distribution? This is because our series of $n = 10$ trials was one sample, with a size of 10, randomly drawn from the population of the infinite number of individual trials we might ever have asked our friend to take. The randomisation came from the way in which we chose the cola to go in each glass and the form of the presentation such that each trial was identical. If our friend had had sufficient stamina, we could have increased the sample size to 20, 100 or 1000 trials. In this case, the population we are generalising to is a rather special one in that it is purely conceptual. No actual population of such trials exists anywhere that we could measure. We can use exactly this logic to deal with the measurement of actually existing populations.

We can use the same logic that we used to work out the probabilities of getting various results from a short series of repeated trials to work out what distribution of results we might get from *any* random sample of trials, of any kind, from *any* population. That includes trials such as measurements, survey questions, administrative records or events and populations comprised of people, objects, institutions, countries or any social phenomena we can measure.

For example, let's switch from our cola experiment to coin flips and deal with the result of getting 8 *heads* in a run of 10 flips. Our sampling distribution would now become one from the population of all coin flips. It would show all the possible results we'd obtain if we repeatedly drew samples of size 10 successive flips from all the flips that could ever exist with a fair coin. The probability of getting at least 8 heads in that series of 10 flips with a fair coin would be 5.5%. That would mean that for every 200 times in the long run, that we carried out a series of 10 flips, we'd come up with 8 or more heads in about 11 of them. Now, instead of coin flips, imagine a population comprising half men and half women. Our sampling distribution would now show the proportion of men we could expect to get if we randomly selected 10 members from it.

We can apply *exactly the same logic* of going from 'the long run' to 'the short run' in a series of trials to going from a population to a random sample taken from that population. When we do this, we use the term *parameter* to refer to a characteristic of the population. Because populations are usually too big or impossible to measure, population parameters are usually *unobserved*. We know that they exist, but we do not know their value because we have not been able to measure them. We use the term *statistic* to refer to our estimate of the value of the population parameter that we make from our sample drawn from the population.

A random sample of 10 people is not very large. Even so, we have seen from the binomial distribution that if it was drawn from a population of any size, no matter how large, in which the probability of picking a woman was 0.5, like correctly guessing the contents of a glass of cola, because the population comprised one half women, then around two thirds of the time, in the long run, we would get a random sample with between 4 and 6 women in it. This corresponds to the 64% of samples of such size we would draw if we repeatedly drew such samples from the population (the distribution was shown in Figure 3.3).

As we increase the sample size, the binomial distribution of k successes in n trials assumes a bell shape that becomes virtually indistinguishable from the Gaussian distribution at around $n = 40$. It follows that for larger samples we can use the Gaussian distribution as our sampling distribution. We can use the fact that we know what proportion of the area under the curve is found within different units of standard deviation from the mean to calculate what proportion of samples we could ever draw that might be found there.

Now, imagine that we take endlessly repeated samples of $n = 100$ from a population that we already know to comprise exactly one half men and one half women, and calculate the statistic for the proportion of women in each sample. What would we find? (We'll deal with the complication that we usually don't know anything about the distribution of men and women in the population later!)

We've already seen from the binomial distribution that the most likely sample to draw would be one in which the proportion of women in the sample was exactly the same as that in the population: 0.5. We can use the same logic as we employed in Chapter 3 to work out the proportion of all the possible sequences of sample results that had different total numbers of women in them from 0 up to 100. However, we could also use a simpler method that gives the same results. If we know that the sampling distribution is Gaussian, all we need to do is calculate its standard deviation. We saw in Chapter 2 that the standard deviation is calculated by summing the squared residuals, dividing by the number of cases and taking the square root. Let's give our women a score of 1 and men a score of 0, so that we have an easy way of tallying the proportion of women in any sample. For any proportion, half the residuals will be the distance from that proportion to 0, and the other half will be the distance from that proportion to 1, so that we get the following formula for the standard deviation (where p is the sample proportion and n the size of the sample):

$$\sqrt{\frac{p*(1-p)}{n}} = \sqrt{\frac{0.5*0.5}{100}} = \sqrt{\frac{0.5*0.5}{10*10}} = \frac{0.5}{10} = 0.05$$

We know from the properties of the Gaussian curve that about 68% of cases lie within 1 SD of the mean. Therefore, in the Gaussian curve, describing the sampling distribution of the proportion of women in every sample size 100 that could ever be drawn, we would find that about 68% of these samples had a proportion of women between $0.5 - 0.05 = 0.45$ and $0.5 + 0.05 = 0.55$ – that is, between 45 and 55 women. We would find about 95% of the samples would have a proportion between $0.5 - (1.96 * 0.05) = 0.402$ and $0.5 + (1.96 * 0.05) = 0.598$. Fractions of women do not exist (I have never met a 0.98 of a woman), so 95% of samples would contain between 40 and 60 women.

Because the bottom line of the formula for the standard deviation is the square root of the number of observations, it is not difficult to see that the amount of variation in the estimate for the proportion of women will depend upon the square root of the sample size. This means that to get half the variation, you need a sample four times bigger. To get 1/10 of the variation, you would need a sample 100 times larger and so on. However, with a sample size as modest as 1000, we would already have a very high probability of getting a very accurate estimate of the proportion of women in the population. The standard deviation of the sampling distribution would now be

$$\sqrt{\frac{0.5*0.5}{1000}} = \sqrt{0.00025} = 0.0158$$

Figure 4.1 illustrates this. More than 95% of samples we drew would contain between 469 and 531 women. We would have almost no probability at all of drawing samples that were much further off the mark, with say, fewer than 400 or more than 600 women.

One final point to emphasise about the sampling distribution. To create it, we do not actually draw any sample at all from the real world. The billions of possible samples in a sampling distribution exist only as mathematical formulas on paper, in our heads or in software. This is why it is so powerful, because if we can work out sampling distributions *theoretically*, by making assumptions about the phenomena we try to measure, that then enables us to compare the empirical results we find in a real sample with what we would have found if our theoretical assumptions had been correct. This turns out to be an incredibly powerful method of analysis, as we shall

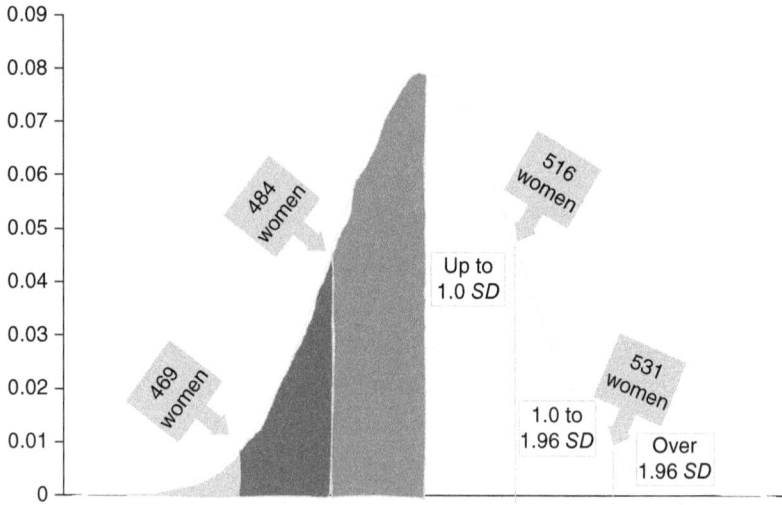

Figure 4.1 The sampling distribution for N = 1000

Note. SD = standard deviation.

see, when we flip our logic, and instead of drawing millions of samples in our imagination, we draw *one* actual sample from the real world. We substitute our theoretical knowledge of the *sampling distribution* for the empirical knowledge of the target population that it would have been impossible to obtain.

Flipping the logic: *one* sample from a distribution

Now, let's use our logic in a slightly different way. Instead of calculating what the distribution of all the possible sample results would look like if we knew the value of the population parameter we were sampling, we can take the estimate we get for the population parameter in *one* sample and instead ask, 'What is the distribution of possible populations that our sample could have come from?' Imagine we have a population of men and women, but we do *not* know what the proportion of men in it is. We draw a random sample of 10 people from it. It contains 8 men. What is our best estimate of the proportion of men in the population? Our sample is the only information we have about the population, and we know from the binomial shape of the sampling distribution that most of the millions of samples of size 10 that we might draw from it will produce an estimate for the proportion of men somewhere near the population proportion. Our best guess will be that men form 80% of the population. We would not expect this guess to be very accurate, but we do have some evidence that there may be more men than women in the population. We could also use the binomial formula, as we did before, to work out a sampling distribution for $p = 0.8$.

It would look like Figure 4.2. Within that histogram, we could see that were the proportion of men in the population actually 0.8, we would expect to get between 7 and 9 men in our 10-person random sample in a proportion of about 0.302 + 0.201 + 0.268 = 0.77 of all the samples that we might ever draw, or just over 75% of our samples.

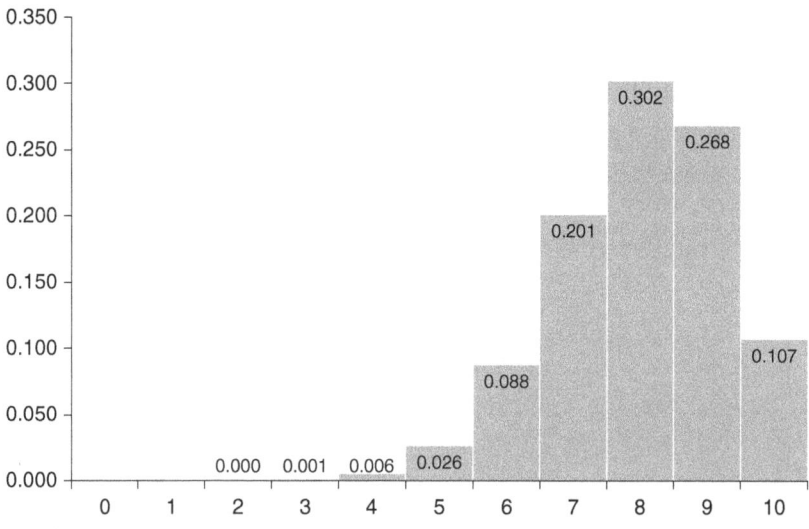

Figure 4.2   The sampling distribution for k when $n = 10$ and $P = 0.8$

A 10-person sample is very small. Let's make it 10 times larger. If I sample 100 people from a population and find 80 men in it, what could I conclude? We can amend the formula we saw earlier for the standard deviation of the binomial distribution to describe the standard error (*SE*) of the sampling distribution for any population proportion:

$$SE = \sqrt{\frac{p*(1-p)}{N}}$$

The standard deviation of a sampling distribution is called the *standard error*. We have seen that the sampling distribution of a statistic appears normally distributed. This is formally demonstrated by the central limit theorem, but we have actually informally established it already. If this is the case, it follows that if we take an area 1.96 *SE* either side of our point estimate of the sample proportion, we will cover the estimate for that proportion that would have been produced by 95% of the samples of this size that we could have drawn. Thus, if we found that the proportion of men in our sample of 100 people was 0.8, we get the following result for the standard error:

$$SE = \sqrt{\frac{0.8*(0.2)}{100}} = \sqrt{\frac{0.16}{100}} = \sqrt{0.0016} = 0.04$$

$$1.96 * 0.04 = 0.078$$

$$0.8 + 0.078 = 0.878$$

$$0.8 - 0.078 = 0.722$$

Adding and subtracting 1.96 *SE* gives us an interval for the proportion of men from 0.722 to 0.878, or from 72% to 88%. This interval is our 95% *confidence interval* (CI) for the proportion of men in the population. We can double check the logic of standard errors and confidence intervals by thinking about the different ways in which we might have ended up with a proportion of 80 men out of 100 in our sample.

There are four possibilities, illustrated in Figure 4.3.

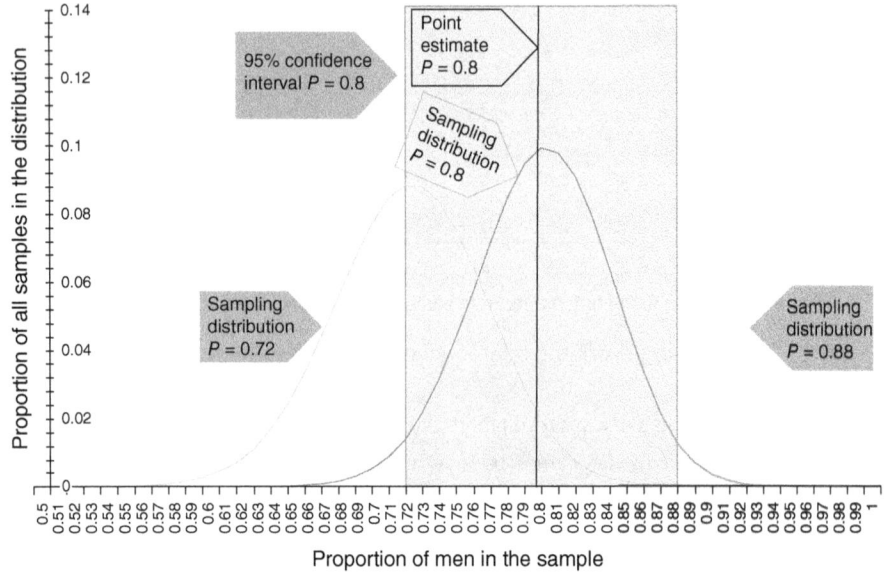

Figure 4.3 Possible sampling distributions for p(male) = 0.8

Our sample could be as follows:

1   The sample with the *highest* proportion of men in it from the 95% of samples nearest the true proportion taken from a population in which 72% are men.
2   The sample with the *lowest* proportion of men in it from the 95% of samples nearest the true proportion taken from a population in which 88% are men.
3   A sample from the 95% of samples nearest the true proportion taken from any population in which more than 72% but less than 88% of the population were men.
4   A sample from the 5% of samples with an estimate *more* than 1.96 *SE* from the population parameter taken *either* from a population in which less than 72% were men *or* from a population with more than 88% men.

Ninety-five percent of the time, our sample will be one of the first three possibilities; 5% of the time it will be the last one. We can therefore say that we are *95% confident* that the proportion of men in the population is between 0.72 and 0.88. Note, however, that there is nothing probable about the proportion of men either in the population or in our sample. Probable men do not exist. What is indeed probable is the status of our knowledge about them. We have used a procedure that will give us a confidence interval which will contain the true population parameter about 19 out of the 20 times that we use it. But for any single occasion when we put this procedure to use, we cannot know which of these two possibilities has occurred.

Once we have drawn our sample and produced our estimate, our confidence interval either does or does not contain the true but unobserved population parameter. This is a matter of definite fact, not probability. But this definite fact is one that we can have no definite knowledge of, since measuring the target population is rarely, if ever, possible. What is a matter of probability is our procedure. Ninety-five percent of the time, it will give us a confidence interval that contains the parameter. Think of randomly selecting an individual from a population that is 95% girls and 5% boys. Before we make the selection, we have a 95% probability of picking a girl. Once we've made it, we have either a boy or a girl, not a youngster that is 95% girl and 5% boy!

Standard errors

Just as we can have a sampling distribution for a proportion, for categorical variables, we can produce a sampling distribution for a mean value for continuous variables. The formula for the standard error is the standard deviation of the population parameter, divided by the square root of the sample size:

$$SE = \frac{\sigma}{\sqrt{n}}$$

where σ, pronounced 'sigma', is the population standard deviation of the variable we are trying to measure. We do not know what this is. Population parameters are unobserved. However, we *can* measure *s*, which is the *sample* standard deviation, and use that instead.

The two standard error formulas, for the mean and the proportion, are actually two versions of the same relationship. Imagine we have a sample size 100 with 50 cats and 50 dogs. The sample proportion of dogs is 0.5. The standard error will be:

$$\frac{\sqrt{p * (1-p)}}{\sqrt{(100-1)}} = \frac{\sqrt{0.5 * (1-0.5)}}{\sqrt{99}} = \frac{\sqrt{0.25}}{9.95} = \frac{0.5}{9.95} = 0.05$$

Let's give the category 'dog' a score of 1 and the category 'cat' a score of 0. What is the mean value of the variable in our sample? It will be the same as the proportion of dogs because:

$$\frac{50*1 + 50*0}{100} = \frac{50}{100} = 0.5$$

What is the sample standard deviation? Every score is either 0.5 above or below the mean, so it must be:

$$\sqrt{\frac{100*0.5*0.5}{100}} = 0.5$$

The standard error must therefore be:

$$\frac{0.5}{\sqrt{100}} = \frac{0.5}{10} = 0.05$$

We can use this formula for standard errors to put confidence intervals around any point estimate drawn from a sample.

Two factors will determine the size of these standard errors: (1) how much variation there is in the phenomenon we are measuring and (2) the size of the sample we use to measure it. Larger samples give smaller standard errors, in proportion to the square root of the sample size. Continuous variables with a large standard deviation – that is to say, with values dispersed far from the mean, or proportions near one half – will produce larger standard errors, as there is more variation to be captured. We can see this by looking at two variables from the study of NCDS children.

Confidence intervals for point estimates

At age 11, the children whose weight at birth we saw in Chapter 2 were given a maths test; 868 girls took the test, getting a mean score of 17.4 out of 40, with an *SD* of 10.5. If we treat these girls as a random sample of girls in the UK aged 11 years in 1969, we could estimate the mean maths ability in the 11-year-old girl population as 17.4 ± 1.96 *SE* to produce a 95% CI:

$$SE = \frac{10.5}{\sqrt{868}} = \frac{10.5}{29.5} = 0.36$$

$$1.96 * 0.36 = 0.70$$

95% CI lower bound for population maths ability = 17.4 – 0.7 = 16.7

95% CI upper bound for population maths ability = 17.4 + 0.7 = 18.1

The height of the girls at age 11 shows much less variation than their maths skill. Their mean height was 57.1 inches with an *SD* of 2.95 inches.

$$SE = \frac{2.95}{\sqrt{807}} = \frac{2.95}{28.4} = 0.104$$

$$1.96 * 0.104 = 0.20$$

We could therefore estimate the mean height of all 11-year-old girls in the UK that year to lie between 56.9 and 57.3 inches.

At that time, 412 out of 853 girls = 0.483 (48.3%) were living in a house that their parents owned or were buying on a mortgage. We could estimate the national proportion as follows:

$$SE = \frac{\sqrt{0.483 * (1 - 0.483)}}{\sqrt{853}} = \frac{\sqrt{0.249}}{29.2} = \frac{0.50}{29.2} \ 0.017$$

$$1.96 * 0.017 = 0.034$$

95% CI lower bound for population proportion = 0.483 − 0.034 = 0.449

95% CI upper bound for population maths ability = 0.483 + 0.034 = 0.517

Thus, we could estimate that nationally between 44.9% and 51.7% of girls aged 11 years were living in houses that their parent(s) owned in 1969 in the UK. At age 23, the women were asked about their highest education qualification. In all, 102 out of 1003 (10.2%) reported having a degree or higher degree. We could therefore estimate the national proportion of women of that age with a degree as follows:

$$SE = \frac{\sqrt{0.102 * (1 - 0.102)}}{\sqrt{1003}} = \frac{\sqrt{0.0916}}{31.7} = \frac{0.303}{31.7} = 0.0095$$

$$1.96 * 0.0095 = 0.019$$

95% CI lower bound for population proportion = 0.102 − 0.019 = 0.083

95% CI upper bound for population maths ability = 0.102 + 0.019 = 0.121

We could estimate that between 8% and 12% of women born in 1958 had a degree or higher degree by 1981. Note that the denominator we use in the calculations is the actual sample size, which will vary from variable to variable in any data set depending on missing responses. Note too that we have to keep in mind our target population. This would not be all women aged 23 years in the UK, but rather women born in the UK in 1958 and still resident there.

Because of the Gaussian shape of the sampling distribution, we can produce confidence intervals for any level of confidence by changing the number of standard errors above and below the mean of the sampling distribution that we use, and thus increase or decrease the proportion of the area of the entire distribution our confidence extends to, and thus our level of confidence. The higher we wish that confidence to be, the farther into the tails of the sampling distribution will we have to reach in order to cover the area where our individual random sample might lie. Because around 95% of the area under a Gaussian distribution lies 1.96 *SD* either side of the mean, ±1.96 *SE* will cover this area. For a 99% probability, we will need to cover 2.54 *SE*. For a 99.9% probability, we will need a little more than 3 *SE*.

The *t*-distribution

In Chapter 2 Section 'Bar charts and histograms', we saw that *z*-scores can be used to express any point on a distribution, or value of a variable, by describing how many standard deviations above or below the mean it lies. Thus, using *z*-scores and standard errors we can move directly between probabilities and intervals. This is a procedure we can use for other methods of statistical inference as well, in analysing contingency tables and regression coefficients.

For very small sample sizes (below 30), the Gaussian distribution and its associated *z*-scores no longer work well because of the way in which the sample standard deviation is used to estimate the population standard deviation to produce an *estimate* of the standard error. (If we knew the population standard deviation, we would be able to calculate the exact standard error but, if we knew the population standard deviation, we would probably also know the values of other parameters in the population and therefore have no need to work out standard errors in the first place!). The *t*-distribution is similar in shape to the Gaussian distribution but with fatter tails, so that larger proportions of the distribution are found beyond any given number of standard deviations from the mean. *T* tables and statistics depend upon the *degrees of freedom (df)* in the data, which is equal to the sample size (*n*) minus 1. Thus, for example, while under the Gaussian distribution about 5% of the area is more than 1.96 *SD* away, for a sample size of 30 (*df* = 30 − 1 = 29) this would become 2.04 *SD*, and for a sample size of 1 of 10 about 2.23 *SD*. Above a sample size of about 30, the *t*-distribution becomes close enough to the Gaussian distribution that the differences can usually be safely ignored. Guinness drinkers will be proud to discover that the *t*-distribution (known as *Student's t*) was discovered by W. S. Gossett who was a statistician and chemist at the Guinness brewery in Dublin. Guinness insisted that he published his articles under a pseudonym lest they inadvertently alert rival brewers to their methods. It is unknown whether he chose the title after watching the drinking habits of Dublin undergraduates.

Examining differences in means and proportions

When they were 23 years old, the NCDS sample were asked about their hourly earning if they were employed. The mean figure for 722 men was £2.69 with an *SD* of 93p, and for 599 women was £2.26 with an *SD* of 78p, so that the mean earnings of men in the sample were 43p per hour greater than the women.

Is this a difference that we could be confident also exists in the population as a whole? One way to tackle this problem would be to look at the confidence intervals for each estimate, using the standard errors and z-scores.

The standard error in pennies for men will be $93/\sqrt{722}$ = 93/26.9 = 3.5p

The standard error in pennies for women will be $78/\sqrt{599}$ = 78/24.5 = 3.2p

For a 95% CI, we'll need to take ±1.96 *SE*

Men £2.69 ± 1.96 * 3.5 ≈ 7p. 95% CI = £2.62 to £2.76

Women £2.26 ± 1.96 * 3.2 ≈ 6p. 95% CI = £2.20 to £2.32

It is clear that women's earnings *were* well below those of their male counterparts, and sampling variation cannot possibly account for the difference.

If the confidence intervals for the two samples (men and women) do not overlap, then that is the same as arguing that the probability of seeing these results, were there no difference in the population, is less than 1 minus the confidence level, in this case 1 – 0.95 = 0.05. Wider 99% CIs would enable us to say that there was a less than 1% chance that women aged 23 years in the population had hourly earnings equal to men. However, if the confidence intervals *do* overlap, that in itself does not indicate that the means for the two groups in the population might be the same.

Let's examine the earnings of men and women who had a degree at age 23. We get the following results:

Men 95% CI £2.72p to £3.00p

(*n* = 158; mean = £2.86; *SD* = 89p; *SE* = 7.1p)

Women 95% CI £2.51 to £2.79

(*n* = 139; mean = £2.65; *SD* = 82p; *SE* = 7.0p)

Our confidence intervals overlap, but what we really want to know is the confidence interval for the *difference in the means* itself, that is the value $\mu_{men} - \mu_{women}$. We therefore need to calculate what proportion of samples in the sampling distribution for the statistic of the difference between the means report a difference of zero between the means. The formula for this involves squaring each standard error, adding them together and taking the square root.

$$SE = \sqrt{\frac{s^2_{men}}{n_{men}} + \frac{s^2_{women}}{n_{women}}} = \sqrt{\frac{0.79}{158} + \frac{0.67}{139}} = \sqrt{0.0098} = 9.9p$$

So our estimate of the difference with a 95% CI is:

£2.86p – £2.65p = 21p ± 1.96 * 9.9p = +1.6p to +40.4p

We have evidence that the difference between men and women in the sample is also one we'd be likely to find in the population, because 19 samples out of 20 in our sampling distribution would give a difference between the means of at least +1.6p. If we wished to be confident of our result at the level of 99%, we could multiply by 2.58 *SE* rather than 1.96. This would give us the following calculation:

£2.86p – £2.65p = 21p ± 2.58 * 9.9p = – 4.5p to +46.5p

Our 99% CI contains the result that the means are indeed equal (a difference of 0.0p), so that we could *not* reject our null hypothesis. This is a useful reminder of the purpose and limits of confidence intervals. What difference is there between 95% and 99% confidence in a result? It is very doubtful that an audience untutored in statistics would distinguish between the two. A rote application of rules would give us a positive result at the 95% level, but not at the 99% level. This illustrates why setting confidence levels *before* conducting an analysis is important: it avoids the temptation to cherry pick the confidence level that delivers a result. It also shows the importance of data exploration as well as formal testing. We have evidence that there is a difference in the earnings of men and women in this cohort, but that evidence is not particularly strong. One reason for this is that our sample is not large. Rather than agonising about whether our evidence is strong enough, it would make sense to look for other data to supplement it. Our sample is limited to a cohort of people aged 23 years in 1981. It would make sense to look for evidence about men and women of different ages, or in other years.

We can use a similar procedure to examine differences in proportions. The NCDS children were asked how they voted in the 1979 general election. Two hundred and fifty out of 656 women (0.381) said they voted Conservative, compared to 258 out of 642 men (0.402). Is this difference in proportion of just over 2% (0.021) evidence of a real difference in the population?

We take the square root of the sum of the squared standard error for each proportion:

$$ SE = \sqrt{\frac{p_1(1-p_1)}{n_1} + \frac{p_2(1-p_2)}{n_2}} = \sqrt{\frac{0.381 * 0.619}{656} + \frac{0.402 * 0.598}{642}} = 0.027 $$

$1.96 * 0.027 = 0.053$

For a 95% CI that would give us a range for the difference from $0.021 - 0.053 = -0.032$ up to $0.021 + 0.053 = +0.074$.

It would *not* be safe to conclude that the small difference we see in our sample is one we would have seen in the population had we been able to measure it. Note that this failure to reject the null hypothesis does *not* allow us to conclude that there *was* no difference between how men and women voted, only that if such a difference existed, its size was not substantial enough for us to be confident about such a result from the evidence we gathered.

What happens to null hypotheses that are not rejected? Rather than being accepted, they settle into a kind of epistemological limbo. This might seem obtuse, and Jerzey Neyman, the original proponent of confidence intervals, made the case for switching the focus onto the costs and benefits of any errors arising from deciding to reject or accept a hypothesis. The most important factor to think about is the *power* of our test, which describes the kind of effect size our methods are capable of discovering. In the example we are looking at here, our sample is not large enough to give solid evidence of a two percentage point difference in voting behaviour, if that is what actually existed in the target population. Had the difference been greater – around 5% – we would have found it easier to detect. Imagine we have found that 235 women (0.358) and 265 men (0.413) had voted Conservative. Now our calculations would be as follows:

$$ SE = \sqrt{\frac{p_1(1-p_1)}{n_1} + \frac{p_2(1-p_2)}{n_2}} = \sqrt{\frac{0.358 * 0.642}{656} + \frac{0.413 * 0.587}{642}} = 0.027, 1.96 * $$

$0.027 = 0.053$

For a 95% CI that would give us a range for the difference from $0.055 - 0.053 = +0.002$ up to $0.055 + 0.053 = +0.108$, suggesting that there was evidence that women were indeed slightly less likely to vote Conservative.

Calculating standard errors for categorical variables

These are calculations that you rarely, if ever, face doing by hand. Statistical software will readily produce standard errors. One exception to this is when producing frequency and contingency tables in software such as IBM SPSS. While the software will report the proportion of cases falling in each category of a variable, it does not report either confidence intervals or standard errors. However, it is always possible to get round this by producing a new version of any variable with the category of interest coded as 1 and other categories set to 0. The standard error for the mean of the variable will now be equal to the standard error of the proportion of that category, as we saw earlier for cats and dogs.

Table 4.1 shows how this can be done. The first four columns show the kind of output that software often produces. It is a contingency table with point estimates to one decimal place of the proportions agreeing and disagreeing with the statement 'Men should have more right to a job than women when jobs are scarce'. In all the countries, a majority of respondents disagreed with the statement, but the proportion agreeing varied from a tiny 1.5% in Iceland to 3 out of 10 in Hungary. This represents a substantial change from when this question was first fielded in surveys in the early 1980s, when *majorities* of respondents in most European countries *agreed* with the statement!

Table 4.1 Response to the statement 'Men should have more right to a job than women when jobs are scarce'

(Row %)	Agree	N.A.N.D.	Disagree	SE (agree)	CI Lower	CI Upper
Iceland	1.5	2.4	96.1	0.0069	0.7	2.3
Sweden	2.1	4.5	93.4	0.0137	1.4	2.8
Finland	3.3	4.9	91.8	0.0248	2.5	4.1
Germany	6.9	15.9	77.2	0.0597	6.0	7.8
United Kingdom	7.0	7.4	85.6	0.0587	5.9	8.1
Spain	7.2	5.6	87.3	0.0602	6.0	8.3
Switzerland	12.9	15.0	72.1	0.1123	11.2	14.6
Portugal	13.1	4.8	82.0	0.1127	11.3	15.0
Austria	14.1	12.4	73.5	0.1256	12.6	15.6
Czech Republic	18.6	19.2	62.2	0.1696	17.0	20.2
Poland	19.6	15.4	65.0	0.1770	17.7	21.5
Italy	21.5	14.6	63.9	0.1989	19.9	23.1
Russian Federation	26.7	22.8	50.5	0.2492	24.9	28.5
HU Hungary	29.6	23.8	46.6	0.2730	27.3	31.8

Note. NAND = neither agree nor disagree; *SE* = standard error; CI = confidence interval.

How much does the picture change when we place confidence intervals round these estimates using standard errors? To do this, a new variable was created coded as 1 for 'agree' and 0 for the other two responses, and the software was asked to calculate the mean and standard error of the mean for this new variable (reported in column 5 of Table 4.1). The mean of the new variable is equal to the percentage in column 2, but expressed as a proportion of 1 rather than as a percentage out of 100. For a 95% CI, 1.96 *SE* of the mean were then added to and subtracted from the mean, and then the result was multiplied by 100 to re-express it as a percentage again. For most countries, the confidence interval spans around 2 to 4 percentage points, as sample sizes in most countries were around 1000 respondents. It would therefore mislead readers about the precision of our original data if we reported point estimates with decimal places. Either the confidence intervals should be reported or the point estimates rounded to the nearest whole percentage point.

When dealing with frequency tables, it is also important not to make misleading summaries of such data that rest upon selectively focusing on the small numbers found at the tail end of a distribution. Someone looking for an attractive quote for a press release might be tempted to make statements like 'Compared to Iceland, people in the UK are over *four times* more likely to agree that men should get priority when jobs are scarce'. However, if we look at the proportions *disagreeing*, they are much closer: 96% and 86%. Put confidence intervals of a couple of percentage points around these two point estimates and the results are closer still. We have good evidence that opposition is stronger in Iceland, but the contrast is not so dramatic.

Producing confidence intervals around point estimates is one of the simplest but also most important procedures in inference. It is good practice to report confidence intervals whenever the data you use comes from a sample (which is almost always the case). This can be done graphically by inserting error bars in bar charts, by producing caterpillar plots or by reporting the confidence intervals whenever point estimates are given.

For example, the NCDS study recorded the type of school the children attended when they were aged 16 years. As we might expect, there was an association between the occupational social class of their father at the time of their birth and the type of school they attended. Figure 4.4 shows a clustered bar chart of school type by class. At the end of each bar is a whisker that represents the 95% CI for the proportion represented by the main bar. Thus, for example, in our data, the point estimate for the proportion of children whose fathers did non-manual jobs who attended grammar schools was 21.5%, but the 95% CI extends around 2 percentage points above and below this.

Figure 4.5 shows a caterpillar plot for the proportion of students at different universities saying that 'overall they are satisfied with their course' in the National Student Survey. Although the percentage of students reporting satisfaction differed across universities, the relatively small numbers of students means that the size of the error bars around the point estimate for each institution means that it is not possible to

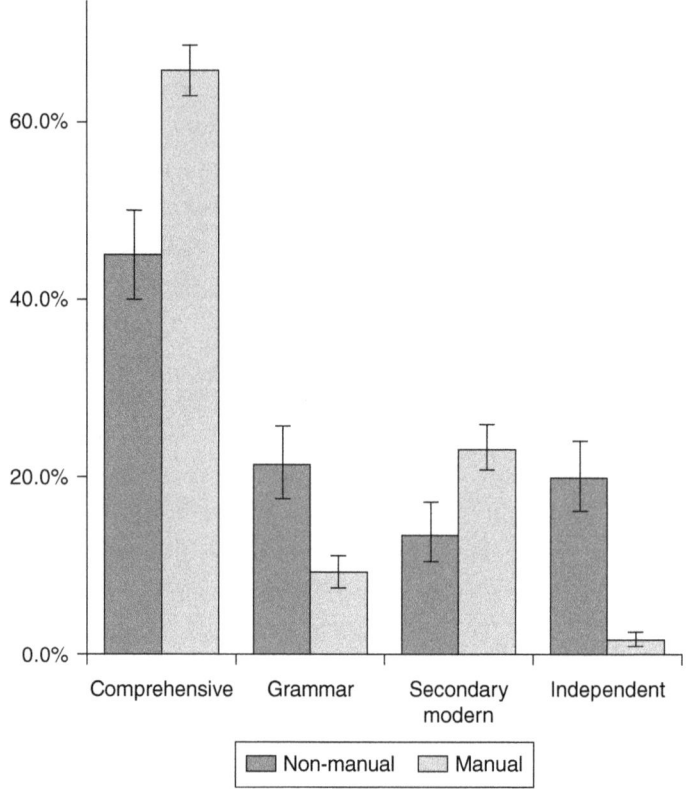

Figure 4.4 Father's social class by type of school at age 16

Note. Data from National Child Development Study teaching data set.

distinguish most universities either from each other or from the average for all universities. Unfortunately, this does not stop university vice chancellors and others from producing misleading 'league tables' of these proportions that really have no basis in the statistics.

So far, we have seen three population parameters and the sampling distributions they give rise to. In Chapter 3, we saw that sampling distribution for the probability of a Bernoulli trial success was given by the binomial distribution of k successes in n trials. At values much beyond $n = 30$, the binomial distribution rapidly converges with that of the Gaussian distribution. The probability of success in a Bernoulli trial can model either a proportion (8 out of 10 glasses of cola correctly identified) or a mean value (a probability of getting 0.8 of glasses tried correct). For the parameters of the population mean and the population proportion, we've seen that the sampling distribution of the sample estimates are Guassian, so that we can calculate standard errors and then use z-scores to multiply these standard errors to produce confidence intervals or p-values.

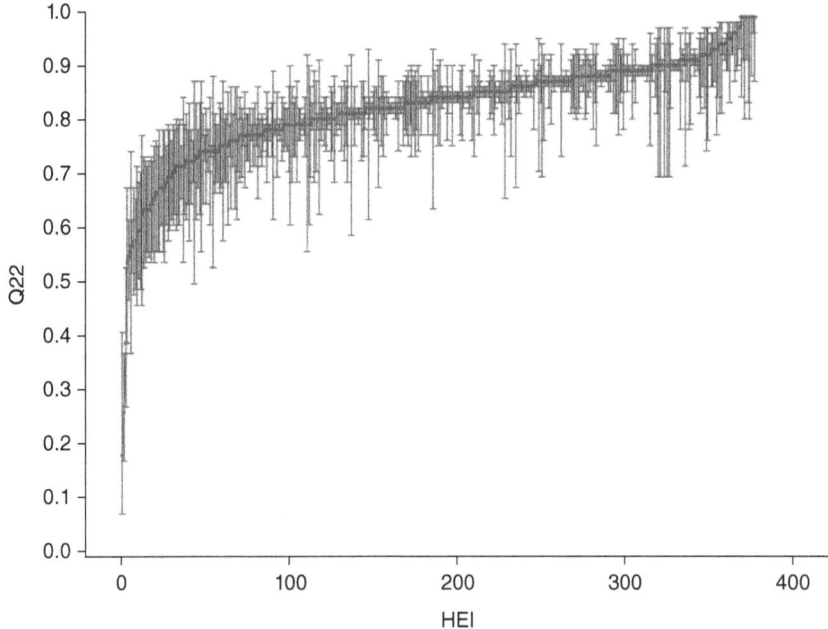

Figure 4.5 Proportion of students 'satisfied' by higher education institutions (National Student Survey)

Note. Data from Office for National Statistics (2016).

The binomial formula in action: stop and search

We have just seen some practical applications of Guassian distribution in making estimates of population parameters based on samples. What about some practical applications of the binomial?

We can use the binomial distribution to answer the question 'How consistent is an *observed* series of n trials with k outcomes of interest with any underlying long run probability of that outcome, or $P(k)$?' Many social and political processes can be modelled as Bernoulli trials in this way. As well as inferring the characteristics of target populations that actually exist, we can also imagine target populations that have never actually existed and sample from them too. At first sight, this might appear to be an odd thing to do, but it turns out to be rather useful. We can do this in order to answer the question 'Could random variation account for any difference that we observe between some actually existing empirical feature of the world and what we might *expect* to see, conditional upon some theory about how that world works?'

For example, an employer who claims to practice gender equality, but in a given year promotes proportionately more male than female staff, could argue that this is only the sort of variation one might expect from year to year, given the small numbers of people involved. We can use the logic of inference to work out the probability that the samples of men and women promoted each year are in effect drawn from the same population so that their sex is irrelevant.

Another example is the stop-and-search behaviour of the police. 'Stop and search' is a police power to stop, question and search a person who is suspected of doing something illegal, such as carrying illegal drugs or an offensive weapon. Critiques of this police power argue that it has been used in a discriminatory way and that black youths are disproportionately targeted for stopping. For the last 10 years, police forces are obliged to record all stop-and-search actions, including the ethnicity of the person stopped. Table 4.2 gives the data for 2017–2018, the last year for which data was available, for the Metropolitan Police, responsible for policing London, and Police Scotland.

Table 4.2   Stop-and-search actions, 2017–2018, Police Scotland and Metropolitan Police

	Metropolitan Police	**Police Scotland**
All searches	116,385	29,805
Searches: black ethnicity *observed*	43,089	384
% of all searches	37.0%	1.29
Population: % black ethnicity	13.3%	1.36
Searches: black ethnicity *expected*	15,479	417
Difference observed – expected	+27,610	−33
Standard deviation	116	20
z-score	+238	−1.65

Note. Police Powers and procedures England and Wales year ending March 2018, Home Office Statistical Bulletin 24/18; Police Scotland Stop and Search Data 1st April 2017–31st March 2018 available at https://www.scotland.police.uk/about-us/police-scotland/stop-and-search/data-publication/;accessed 22/12/2020

Just as with samples and populations, we can establish a null hypothesis, and calculate the probability distribution of the data we would *expect* to see if the null were true. We can then calculate the probability of obtaining the data we observe were the null true. An appropriate null hypothesis here is that the ethnic breakdown of those stopped and searched is similar to that of the population in the area that is policed. The population in London identifying as black is about 10 times the percentage in Scotland. London has long been a cosmopolitan city drawing in global migration so that around one third of its population was born outwith the UK. By contrast, the Scottish economy has been weaker than that of England for much of the last half-century, so that it has tended to be a net exporter of people and has drawn in fewer immigrants of Caribbean or African descent. We can see that in Scotland police actually stopped and searched fewer people of this ethnicity than their proportion in the population. However, at only a little more than 1.5 SD from the expected number of searches, it is within the range one might expect of random variation from year to year. The situation in London could not be more different. The number stopped and searched is a staggering 238 SD above the expected value. It is inconceivable that such a discrepancy could have occurred by chance. However, we still need to consider our null hypothesis. Is it reasonable? While we might expect the proportion of stops by ethnicity to be broadly in line with the ethnic breakdown of the population, there could be a number of factors that account for some of the difference. For example, public law enforcement often focuses on young men. The proportion of young black males in the population may be a little higher than for the population as a whole. The location of residence and where the stop takes place may not coincide. It is hard to think of any factor that would allow us to avoid the inference from our data that the Metropolitan Police, either consciously or not, deliberately targets people who identify as black in their stop-and-search procedures.

Chi-Square and contingency tables

Analysis in the social sciences often takes the form of contingency tables showing the probability distribution of one categorical variable conditional upon the values of another categorical variable. Such tables are common because they are excellent at showing an association between variables in an intuitive way that audiences can readily understand because, unlike coefficients or measures of association, the raw frequencies are visible in the table.

The relationship between father's occupational social class and type of school children attended at age 16 that we saw earlier in Figure 4.4 is shown as a contingency table in Table 4.3. You may find it helpful to think of the contingency table as the probability distributions of the results of *two* trials undertaken by each child in the study:

- What type of school was attended at year 16?
- What occupation did the father have at the child's birth?

We want to know if these two trials turn out to be independent of each other, or whether there is some association, and whether any evidence of association we do find is strong enough to enable us to infer that this association also existed in the population from which our sample was drawn.

Table 4.3 Father's class by type of school

(Column %)	Non-Manual	Manual
Comprehensive	45.0	65.9
Secondary modern	13.6	23.4
Grammar	21.5	9.1
Independent	19.9	1.6
All	100.0	100.0
N	382	1006

Note. Data from National Child Development Study teaching data set.

As before, we can form a null hypothesis, use it to produce a sampling distribution and then examine the difference between what we would *expect* to see under a null hypothesis of no association and what we actually do observe. If the probability of observing the data we have in our empirical table, conditional upon it being drawn from the sampling distribution for the null hypothesis, is low enough, we can provisionally conclude, with the appropriate level of confidence, that an association at least as strong as that which we observe in the table *also* exists in the target population from which our sample was drawn. The sampling distribution will be that of the χ^2 (chi-square, pronounced 'ky' as in 'sky') statistic for the table. Where does this come from?

I suspect that most people who ask software to produce tests, or readers of papers reporting the results of them, have little idea what χ^2 is. This is unfortunate because understanding χ^2 is not very demanding and provides the best insurance against its abuse. χ^2 is no more mysterious than the binomial, but it is more general, and thus more powerful.

We've already met the Guassian distribution, illustrated empirically by Union soldiers and NCDS babies and theoretically by the standard normal distribution with its distinctive bell shape. Recall that we can standardise this distribution by re-expressing its values as multiples of 1 *SD* from the mean or z-scores. Figure 4.6 is a reminder of such a distribution. Most observations are near the mean (z = 0) and become less frequent as we move further from it, or as z increases. As with any probability distribution, we can treat the total area under the curve as one and the proportion of the area above any two points on the horizontal axis of the graph as equal to the probability of finding an observation within that interval of values. The two shaded areas representing the 0.025 + 0.025 = 0.05 of the total area are located just under 2 *SDs* (z = ±1.96) from the mean in each 'tail' of the distribution. If this were a graph of the height of Union soldiers, it would correspond to soldiers who were either below 5 ft 1½ in. or above 6ft ½ in. in height as we saw in Chapter 2 Section 'Height of Union soldiers'.

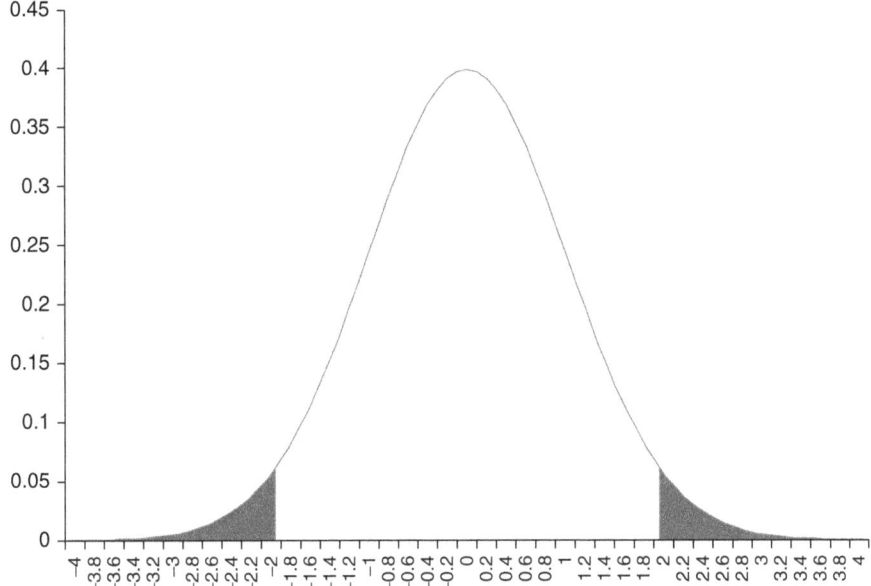

Figure 4.6 A standard normal distribution

We could treat these z-scores as observed *deviations* from an *expected* value equal to the mean of the distribution – that is, as *residuals* from it. The mean height of Union soldiers was 5 ft 7 in. We would expect most soldiers to be fairly near this height. As we go further away from this *expected* height, as we deviate farther from it or as the size of the residual increases, so does the probability of any *observed* height decrease. If we pick a Union soldier at random, we'd have only a 5% chance of getting one shorter than 5 ft 1½ in. or taller than 6 ft ½ in. The z-scores are just a logically robust way of describing how the probability of observing any value declines as the size of the residual increases.

Just as we did when calculating the standard deviation, we could square the value of these residuals to get rid of the sign, so that our attention is only on how far we are away from the mean of the distribution and not on whether we are above or below it. When we produce a chart of these squared residuals, the horizontal axis will now start at zero, and its height will be double our previous chart. This is shown in Figure 4.7. It is as if we had folded Figure 4.6 onto itself.

As before, the shaded area will represent $0.025 + 0.025 = 0.05$ of the total area of the graph, but it will now be located exclusively in the right-hand tail of the graph, corresponding to the area to the right of $1.96^2 = 3.84$ squared residuals or z-scores. You are looking at the distribution of χ^2 for 1 *degree of freedom*. The shape of the graph is what we ought to expect. In an approximately normal distribution, most values

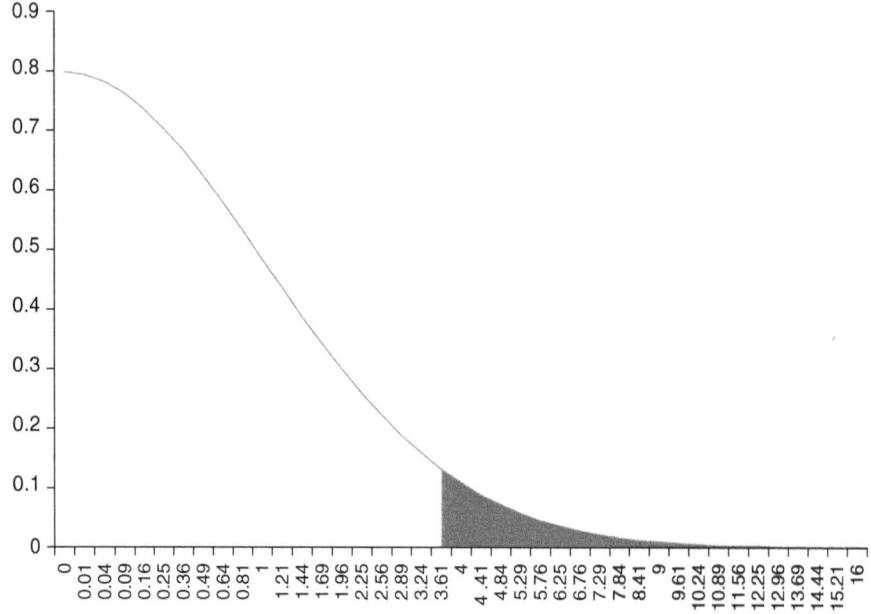

Figure 4.7 Squared residuals of a standard normal distribution

will be close to the mean, so that the value of their squared deviations will be close to zero. As we move farther from the mean, and the value of the squared deviations increases, the probability of an observation falls away.

Consider what would happen if we had more than one standard normalised deviate variable? Alongside our first variable (squared deviates from mean height) we might have squared deviates from mean age. If we summed these square deviates for each observation, what would happen to the shape of our distribution? It would shift rightward, since we now have two amounts of squared deviation for each case. It would also become more spread out and the height of the curve corresponding to zero deviation would fall, since this would now correspond only to soldiers who were exactly of mean height *and* of mean age. Each successive variable added would shift our curve downward and rightward. As we added more variables, the proportion of observations close to the mean for *all* the variables would quickly fall towards zero, and the 'hump' in our curve resulting from this fall would shift rightward. Eventually, as the number of variables was increased, the χ^2 distribution would approximate a normal distribution. Recall that in Chapter 2 we saw that an approximately normal curve results from the summation of several random variables. Most scores will be a mix of above-average and below-average scores on individual variables and therefore close to the grand mean. Progressively fewer observations will comprise scores that are consistently above or below average, leading to the familiar bell shape. Figure 4.8 shows the χ^2 distributions obtained for one through nine variables, or *degrees of freedom*.

The concept of degrees of freedom is explained in the appendix to this chapter. The key point to grasp is that it describes the number of components of a model or equation that are not fixed but free to vary – in this case, the number of standard normalised variables we are considering. In statistical notation, a subscript is used to indicate the degrees of freedom: χ_k^2 would refer to k degrees of freedom.

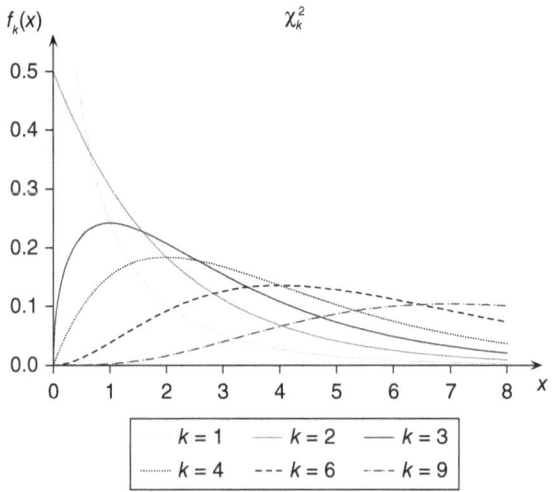

Figure 4.8   Chi-Square distributions for one to nine degrees of freedom

Note. Geek 3 at https://commons.wikimedia.org/w/index.php?curid=9884213

Let's return to our contingency table of father's class and school type armed with our knowledge of χ^2 and proceed to construct a null hypothesis. Table 4.4 records the distribution of the values *observed* for the children for these two variables as raw counts.

Table 4.4   Father's class by type of school: observed frequencies

Observed	Non-Manual	Manual	All
Comprehensive	172	663	835
Secondary modern	52	235	287
Grammar	82	92	174
Independent	76	16	92
N	382	1006	1388

Note. Data from National Child Development Study teaching data set.

Using the table margins, we can calculate the numbers of children we would *expect* to see in each cell of the table were there no association. Notice how these expected values leave the marginal values in the table unchanged. That is, it leaves the original marginal probability distributions of the two trials exactly as they were, but it rearranges the cell contents such that as we go down the columns, the row percentages do not change, and as we move across the rows, the column percentages do not change. This is shown in Table 4.5.

Table 4.5 Father's class by type of school: expected frequencies

Expected	Non-Manual	Manual	All
Comprehensive	835*382/1388=230	835*1006/1388=605	835
Secondary modern	287*382= 79	287*1006/1388= 208	287
Grammar	174*382/1388= 48	174*1006/1388=126	174
Independent	92*382/1388=25	92*1006/1388=67	92
N	382	1006	1388

Note. Data from National Child Development Study teaching data set.

As usual, we find a residual by subtracting our expected value from the observed value (as shown in Table 4.6), we then square the residual to get rid of the sign, and finally we divide by the number of expected cases to standardise the residual – that is, leave it unaffected by the total number of our observations (Table 4.7).

Table 4.6 Residuals: observed–expected

Residuals: observed-expected	Non-Manual	Manual	Total
Comprehensive	−58.0	58.0	0
Secondary modern	−27.0	27.0	0
Grammar	34.0	−34.0	0
Independent	51.0	−51.0	0
Total	0	0	0

Note. Data from National Child Development Study teaching data set.

Table 4.7 (Residuals squared)/expected

	Non-Manual	Manual	All
Comprehensive	14.6	5.6	20.2
Secondary modern	9.2	3.5	12.7
Grammar	24.1	9.2	33.3
Independent	104.0	38.8	142.9
N	152.0	57.1	209.0

Note. Data from National Child Development Study teaching data set.

If we now sum these residuals, the total is 209:

$$\chi^2 = 14.6 + 5.6 + 9.2 + 3.5 + 24.1 + 9.2 + 104.0 + 38.8 = 209.0$$

We now need to work out the probability of observing a total difference between the observed and the expected cell counts this large when the null hypothesis of no association is true. This requires the χ^2 distribution. But how many degrees of freedom do we have? This is not the same as the number of original variables in the table, but rather the total number of cell counts that are free to vary when the table marginal counts are fixed, because these are the random quantities that we add up to arrive at our value for the χ^2 statistic. This is the same as asking the question 'What is the number of cells whose value I have to fix before all the other cell totals can be filled in by using the marginals?'

For a simple table such as ours, we can work this out empirically. Let's start with the top left cell, corresponding to children of non-manual fathers going to comprehensive schools. In order to fit with the existing table marginal totals for this cell (382 and 835), I could enter any number between 0 and 382, so that the amount in this cell is free to vary. However, once I define a value for this cell, I automatically define a value for the cell in the next column. It is not free to vary, since it *must* take the value of what I entered in the first cell subtracted from the marginal row total of 835. If I move down to the cell for non-manual fathered children attending secondary modern schools, I have a number that is free to vary once more. It can take any value between 0 and 382 minus the value of the cell above. The same will go for the cell below too. For each of these cells, however, once I fix their value, the value of the other cell in the row will be automatically defined. However, when I come to the last cell in this column, I will discover that it must already be defined if I have fixed a value for the first three cells in the column. It has to take the value of the column marginal (382) minus the total of three cells above it. Thus, in the eight cells in my table, if I fix the value of any three cells, I can fill in the values of all the others from the table marginals. I therefore have 3 *df*. The formula for the number of degrees of freedom in any two-way contingency table is

$$df = (N \text{ rows} - 1) * (N \text{ columns} - 1)$$

So for our table, here it is

$$df = (4 - 1) * (2 - 1) = 3$$

You can easily find χ^2 calculators online, but usually whatever software you use to produce your contingency table will also calculate χ^2 for you. Figure 4.9 shows the sampling distribution of the χ^2 statistic for 3 *df*. The probability of us observing the data in our table if the null hypothesis of no association were true is vanishingly small. Even a value 1/10 that of what we found would have been strong evidence of a relationship. However, demonstrating that we can reject the null, and that we have evidence of association, does not exhaust what we can say.

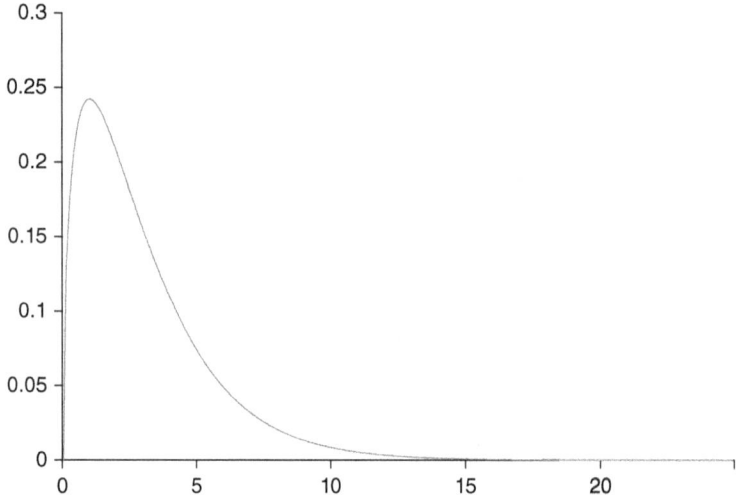

Figure 4.9 Distribution, *df* = 3

Note. df = degrees of freedom.

We can look at our table and the associated x^2 value for each cell. This would allow us to make a comment about the importance of independent schools in influencing the different types of school that children attended. We might also draw attention to how father's social class influenced different probabilities of going to grammar schools. However, we cannot simply conclude from this table that 'fathers social class determines the type of school children attend', for four reasons. First, it is only one of many influences, albeit a substantial one. Almost 6 out of 10 children with non-manual fathers went to comprehensive or secondary modern schools, and about 1 in 10 children of manual fathers nevertheless got to grammar or independent schools. Second, our data is for one cohort of children from 1 year of birth, attending schools in the same time period. Other cohorts of children may have had different experiences. Third, we need to consider measurement error. Some children did not take part in the study or could not be followed up, or respondents may not have answered questions or answered them wrongly, or interviewers might have misinterpreted their answers or the way in which occupations were assigned to the two classes we described may have been open to debate. No matter how expertly we measure father's class and children's school, these can change over time. I find it useful, as a rough rule of thumb, to put an interval of plus or minus 1/10 of the value in each contingency table cell before being confident that an association strong enough to allow me to reject the null hypothesis is also one that is a signal substantial enough to have arisen through the unknown amount of measurement noise. Applying this rule of thumb to Table 4.3 would leave the story it tells unchanged, as shown in Table 4.8.

Table 4.8 Father's class by type of school

	Non-Manual (%)	Manual (%)
Comprehensive	41–50	60–72
Secondary modern	12–15	21–26
Grammar	19–24	8–10
Independent	18–22	1.4–1.8
All	100.0	100.0
N	382	1006

Note. Data from National Child Development Study teaching data set.

Chi-Square and goodness of fit

We could use the x^2 distribution to solve the problem of the probability of getting 8 out of 10 on the Coke and Pepsi test. We have 1 df (the only thing that can vary is the number of correct identifications). Our observed value is 8 and our expected value was 5. We need to know the standard deviation of our expected value (which we get from the binomial formula, which is $\sqrt{n*p*(1-p)} = \sqrt{10*0.5*0.5} = 1.58$). The standardised value of our observed value is therefore (8 − 5)/1.58 = 1.90. We square this to obtain the value 3.6. This is the value of x_1^2. How far along the distribution do we find this value? We can find the result from tables, or ask any statistical software to provide it. Just by eyeballing Figure 4.7, you can see that the proportion of values of x_1^2 falling to the right of 3.6 is small – a little above the 5% we know lies to the right of 3.84. The precise value is *exactly the same as we obtained using the binomial formula* because the calculations are actually the same!

The beauty of x^2 is that it allows us to do this for any situation where we wish to make an estimate of how likely *any* empirically observed value is to be observed, were it to have come from some distribution that we expect. The distribution that we expect is described by our null hypothesis. We therefore can use x^2 to see how well the distribution we empirically observe *fits* with that *expected* distribution. This is sometimes referred to as a *goodness-of-fit* test. With the generalised formula that x^2 gives us, we can now create null hypotheses with as many components as we like. All we need to do is standardise the values of the variables involved and work out the number of degrees of freedom. We can leave such calculations to the software, as long as we understand what we are doing.

In Chapter 1, we saw the results from the Salk vaccine trial. One hundred and forty-two unvaccinated children contracted polio, compared to 57 who had been vaccinated. We can use x^2 to work out the probability of getting this data, under the condition that the null hypothesis that the vaccine did *not* work was true. Under the null hypothesis, if the vaccine was worthless, we would expect to see similar numbers of children in each group. If the number we *expect* to see under the null is 99.5 children in each group, what is the probability of random variation accounting for the difference between this number and the number we observe (Table 4.9):

Table 4.9   Salk vaccine trial results

	O	E	O – E	(O – E)²/E
Vaccinated	57	99.5	−42.5	18.15
Unvaccinated	142	99.5	+42.5	18.15

Note. O = observed; E = expected.

The value for x^2 is 18.15 + 18.15 = 36.3, with 1 *df* (if we know the total number of polio cases, any change in the number of vaccinated contracting polio would change the number of those not contracting it). If we take the square root of this value, we'll get the number of standard deviations along the standard normal distribution from the mean. The area under the curve 6 *SD* from the mean is around $p = 0.000000001$, or one in one billion.

John Tukey, the founder of Exploratory Data Analysis, was fond of quipping that statistics was often very good at giving a precise answer to the wrong question. x^2 distributions can often be good at doing just this. It does not mean that they have no value. On the contrary, they are an indispensable means of summarising the difference between any pattern of observed and expected data. The challenge lies in a good definition of 'expected' and a good understanding of what is observed. Let's work through a x^2 test for three contingency tables, based on European Social Survey data from three separate countries on attitudes to homosexuality. They are for Italy, France and Spain. Table 4.10 shows the distribution of support for each option in each of the three countries by sex. To show how x^2 is calculated, I have shown the working for each table cell, the total value for x^2 and the probability associated with that value for (5 − 1) ∗ (2 − 1) = 4 *df* (x_4^2) in Table 4.11.

Table 4.10 (Dis)agreement with the statement 'Gays and lesbians should be free to live life as they wish': Italy, France and Spain, 2016

	Italy (%)			France (%)			Spain (%)		
	Male	Female	All	Male	Female	All	Male	Female	All
Agree strongly	23	25	24	63	68	65	51	60	55
Agree	43	43	43	25	21	23	37	31	34
N.A.N.D.	18	18	18	5	6	6	8	5	6
Disagree	9	9	9	3	2	3	3	2	3
Disagree strongly	7	5	6	4	3	4	1	2	2
	100	100	100	100	100	100	100	100	100

Note. NAND = neither agree nor disagree. Data from European Social Survey Round 8.

Table 4.11 Calculations

		Male	Female	O – E		(O – E²)		(O – E)²/E	
Italy									
Agree strongly	O	285	323	−14.0	13.8	190.4	190.4	0.64	0.62
	E	298.8	309.2						
Agree	O	532	545	2.7	−2.7	7.3	7.3	0.01	0.01
	E	529.3	547.7						
N.A.N.D.	O	223	234	−1.6	1.6	2.6	2.6	0.01	0.01
	E	224.6	232.4						
Disagree	O	115	116	1.5	−1.5	2.3	2.3	0.02	0.02
	E	113.5	117.5						
Disagree strongly	O	81	61	11.2	−11.2	125.4	125.4	1.80	1.74
	E	69.8	72.2						
All	O	1236	1279					4.88	$p = 0.30$
France									
Agree strongly	O	588	751	−26.3	26.3	691.69	691.69	1.13	0.95
	E	614.3	724.7						
Agree	O	239	228	24.7	−24.7	610.09	610.09	2.85	2.41
	E	214.3	252.7						
N.A.N.D.	O	50	68	−4.1	4.1	16.81	16.81	0.31	0.26
	E	54.1	63.9						
Disagree	O	30	26	4.3	−4.3	18.49	18.49	0.72	0.61
	E	25.7	30.3						
Disagree strongly	O	34	37	1.4	−1.4	1.96	1.96	0.06	0.05
	E	32.6	38.4						
All	O	941	1110					9.36	$p = 0.05$

(Continued)

Table 4.11 (Continued)

		Male	Female	O – E		(O – E²)		(O – E)²/E	
Spain									
Agree strongly	O	489	566	−40.2	40	1616.04	1616.04	3.05	3.07
	E	529.2	525.8						
Agree	O	352	297	26.5	−26.5	702.25	702.25	2.16	2.17
	E	325.5	323.5						
N.A.N.D.	O	74	47	13.3	−13.3	176.89	176.89	2.91	2.93
	E	60.7	60.3						
Disagree	O	30	23	3.4	−3.4	11.56	11.56	0.43	0.44
	E	26.6	26.4						
Disagree strongly	O	11	17	−3	3	9	9	0.64	0.64
	E	14	14						
All	O	956	950					18.46	$p = 0.01$

Note. O = observed; E = expected; NAND = neither agree nor disagree.

Looking first at our summary table, how might we describe the results? In all three countries, there is little disagreement with the proposition. There is slightly more in Italy, and here there are also more people who say they 'neither agree nor disagree'; however even in Italy, well over four times as many people agree with the statement as disagree with it. What about differences between men and women in their attitudes? In Italy, there is no difference at all. In France and Spain, women are very slightly more likely to agree than men, but the difference mostly concerns the distribution of responses between the agree and the strongly agree responses. In each country, women are slightly more likely than men to describe their responses as 'strongly agree' rather than 'agree'. How much weight might we put on such a distinction? My overall judgement is that we have good evidence here that there is no difference, and certainly no substantial difference, between the views of men and women on the issue. I could also describe this as the distribution of views on homosexuality being independent of sex in these countries.

What happens if I do a χ^2 goodness-of-fit test of these observed values, against the values I'd expect to see if there were indeed no association between these two variables? The results for Italy clearly do *not* allow us to reject a null hypothesis. Moreover, because of the substantial sample size, it is very unlikely that our test is underpowered, or that the reason for the null standing is only because we have not collected enough evidence. We *could* reject the null with 95% confidence for France and 99% confidence for Spain. However, it would be highly misleading to conclude, because of this, that there is a difference between men and women on this issue in these two countries. There is a *statistically significant* difference in the pattern of views between men and women in these two countries, but it is caused only by a difference in the proportion of men and women who *strongly* agree as opposed to agree with the statement. In both Spain and France, women are slightly more likely to strongly agree. Is this a *substantively* important difference? Almost certainly not. We would need to be sure both that all respondents had a clear and consistent definition of the distinction between agreement and strong agreement, and that this difference had implications for other views or behaviours. I doubt if either of these propositions is true. The moral of the story is clear. Statistical significance need not imply substantive importance. Statistical significance only signals a result that *may* be worth further investigation. It is only one *part* of the logic of inference, not the whole process.

Fisher's exact test

The x^2 formula works well for large numbers, but when the number of observed or expected cases in a cell drops below 5 it can give less reliable results. To get round this, it is possible to use *Fisher's exact test* for 2 × 2 contingency tables. This consists of using a variant of the binomial formula to work out the marginal probability and then calculate the exact probability of obtaining the observed values, or values more extreme.

Jospeh Lister pioneered infection control to sterilise wounds in surgery at Glasgow Royal Infirmary. As Newsom (2003) describes, before the introduction of this technique, in the years 1864 to 1866, 35 patients had limbs amputated, and almost half (16) died as infection set in after the operation. But between 1867 and 1869, using 'German Creosote' containing carbolic acid to coat the wound after amputation, only 6 out of 40 patients succumbed (Table 4.12). With hindsight, Lister's discovery appears obvious. But contemporaries were sceptical or antagonistic, including hospital authorities who were faced with the prospect of having to provide a sterile environment for treatment. What might statistics have made of his discovery? The results of Fisher's exact test are set out below.

Table 4.12 Patient survival after limb amputation

	Died	Survived	All
1864–1866	16	19	35
1867–1869	6	34	40
All	22	53	75

P(died) = 22/75 = 0.293

Note that the table only has 1 *df*, because if we fix the marginal frequencies as they are, specifying the value of any cell in the body of the table also defines what the other values *must* be to keep the marginals unchanged. Let's look at the bottom left cell with observed value 6. It could take any value from 0 up to 22 without changing any marginals. We want to estimate the probability that it would take a value equal to 6 or less, out of 22, were there no association between the two variables. If Lister's treatment made no difference, there ought to be the same underlying probability of death in each time period. This is our null hypothesis. We therefore want to know the probability of getting *x* out of 22 deaths after the treatment was applied, if 40 out of the 75 patients received this treatment. The relevant probability distribution is a variant of the binomial, which goes by the unwieldy title of the hypergeometric distribution. While the binomial distribution assumes a constant long-term probability rate for the trial outcome, the hypergeometric distribution uses the empirical probabilities of the trial itself under the two conditions. Thus, it calculates the probability of *x* (6) from *n* (22) trials, from a population of size *m* (75) in which *k* (40) received the treatment. (There is a function in Excel which allows you to do this). Table 4.13 and Figure 4.10 show the sampling distribution for the null hypothesis of the treatment having no effect using this approach. The cumulative probability of observing as few as six deaths is highlighted in bold in Table 4.13 and is very small: about one third of 1%. Most software packages will conduct *Fisher's exact test* as an alternative to χ^2 where small cell counts are a problem.

Table 4.13 Table P(deaths | H_0 = true)

	P	Cumulative P
0	0.0000	0.0000
1	0.0000	0.0000
2	0.0000	0.0000
3	0.0000	0.0000
4	0.0001	0.0001
5	0.0006	0.0007
6	0.0030	0.0037
7	0.0117	0.0154
8	0.0346	0.0500
9	0.0782	0.1282
10	0.1370	0.2651
11	0.1868	0.4520
12	0.1986	0.6506
13	0.1646	0.8152
14	0.1058	0.9209
15	0.0524	0.9733

	P	Cumulative P
16	0.0198	0.9931
17	0.0056	0.9987
18	0.0011	0.9998
19	0.0002	1.0000
20	0.0000	1.0000
21	0.0000	1.0000
22	0.0000	1.0000

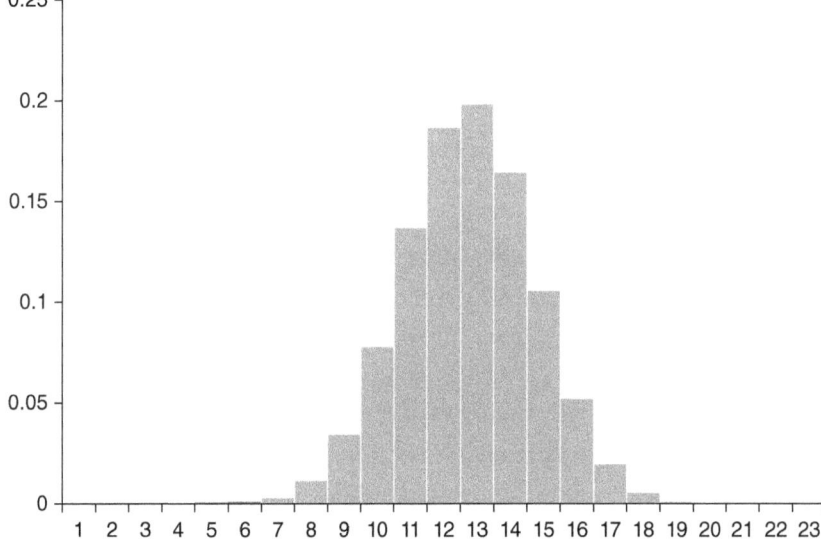

Figure 4.10 Sampling distribution for Fisher's exact test

How does our test result compare with the probability estimated by x^2 ? Remember our formula:

$$x^2 = \sum \frac{(0-E)^2}{E}$$

We can use the table marginals to work out the frequencies *expected* were the null hypothesis true (Table 4.14):

Table 4.14 Expected death and survival under null hypothesis

	Died	Survived	All
1864–1866	22 * 35/75 = 10.3	53 * 35/75= 24.7	35
1867–1869	22 * 40/75= 11.7	53 * 40/75= 28.3	40
All	22	53	75

We can then take the difference between the observed and the expected frequencies in each cell, square it and divide it by the expected frequencies (Table 4.15)

Table 4.15 Calculating chi-square for null hypothesis

	Died	Survived	All
1864–1866	3.20	1.33	35
1867–1869	2.80	1.16	40
All	22	53	75

Summing these gives us a total of 8.5 for x^2, with 1 *df*, which corresponds to a probability of 0.00356 – very close to the exact probability of the Fisher's test.

Appendix: Degrees of freedom

Degrees of freedom is a concept widely used in statistical analysis, often by people who have no clear idea of what it is, unless they have some background in mathematics. This is unfortunate as the concept is not too hard to grasp. There is an excellent introduction to it, well worth reading, published 80 years ago in a psychology journal by Helen Walker, who went on to become the first woman to be elected President of the American Statistical Association (Walker, 1940). The exposition here follows that article.

It is useful to first think about geometry and space. A point on a line exists in only one dimension. It can move forwards and backwards along that line, but has no other freedom of movement. This would be the case even if the line sat on a plane of two dimensions. The line might snake around in any direction on the plane, like a railway track, yet as long as it is confined to the line, our point would have only one way in which it could move. We could determine its position on the line by a single number, for example, the length along the line from its origin. We could describe this situation by saying that it had 1 *df* only. A point on a two-dimensional plane has 2 *df*, and similarly, it would require two numbers or coordinates to describe its position (e.g. horizontal and vertical distance from an origin, or distance from an origin and angle). Thus, the position on a map could be defined by longitude and latitude. That plane described by such a map could in turn be seen as a section through three-dimensional space; nevertheless, any coordinate on that map would have only 2 *df*. If we wished to locate an aeroplane in flight, we would need three coordinates – altitude, longitude and latitude. Were we to track its movement over time, we would need to consider these three dimensions as lying upon a fourth one, of time, so that we would have 4 *df*.

We can represent such geometry with algebra, using equations to describe points, lines or planes in space. The equation $x + y = 5$ would define the line reproduced in Figure 4A.1.

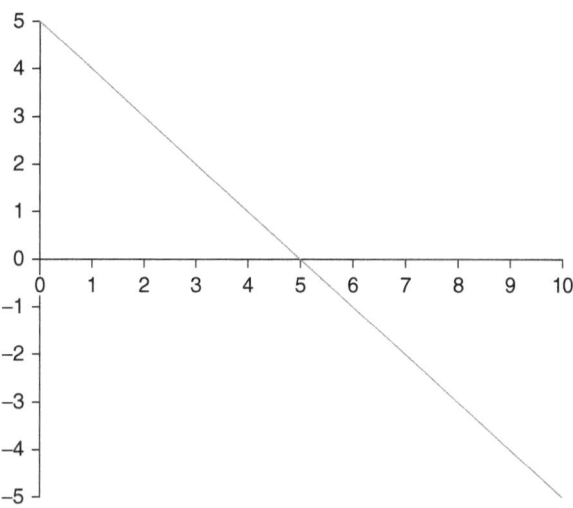

Figure 4A.1 $X + Y = 5$

Such a line would have only 1 *df*, since whatever value we choose for either X or Y leaves us with only one other value that makes their sum equal to 5. The equation $X + Y + Z = 5$ would define a two-dimensional plane in a three-dimensional space. It would have 2 *df*. We can satisfy the terms of the equation with any two values for two of its terms, but to maintain a sum equal to 5, the third number would be fixed. Our imaginations can cope well with only three dimensions, but mathematically, we can envisage as many dimensions as we like. Any set of N numbers can be used to define a single point in N-dimensional space, with N *df*. Specifying a relationship between these numbers will constrain the value that can be taken by one of them, which will reduce the number of degrees of freedom by one. The generalisation to all kinds of statistical analysis requires one final effort of imagination. Our set of N numbers could be the sample of observed values for a random variable. This sample too could be represented as a single point in N-dimensional space with each of the numbers specifying one of its N coordinates. Having N *df*, this point would be free to move anywhere within this N-dimensional space. If we then specified some relationship within this set of coordinates, for example, fixing their mean value, then this would reduce by one the number of dimensions through which the point might move.

Chapter Summary

Like Chapter 3, this chapter has covered a lot of ground. If you have followed the argument carefully, you should be able to understand the following key points:

- When we cannot measure a population, the only sample from it that we can expect to be representative of the population is one that is randomly drawn.
- Random samples that are large enough (above about $n = 30$) are drawn from sampling distributions that are approximately normal in shape with a standard deviation, known as a standard error, that varies with the square root of the sample size and the variance of the characteristic we are trying to estimate.
- We can use standard errors to construct confidence intervals that describe the degree of uncertainty that we have about any point estimates we make using our sample data.
- We can calculate the probability of observing any pattern, result or association we see in our sample data, conditional upon our sample being drawn from a population in which our null hypothesis holds. If the probability of this is small enough, we have provisional evidence to suggest that our sample data came from some other population, one *not* described by the null hypothesis.
- The χ^2 distribution for the relevant number of degrees of freedom is a versatile measure of 'goodness of fit'.
- Fisher's exact test can be used when very small cell sizes render χ^2 unreliable.

A good check on how well you have followed this chapter is to examine Figure 4.3. If you understand its contents, you have understood this chapter.

Further Reading

Any introductory statistics textbook will have a section on inference from samples to population, tests of significance and analysis of categorical data, and any search of the internet for these terms throws up dozens of lecture outlines and PowerPoints. What one person thinks is a clear and accessible summary is gobbledygook to the next, so that it is worthwhile dipping into different treatments to find one that makes sense to you. A clear and short resume of the main arguments is provided in Driscoll, P., Lecky, F., & Crosby, M. (2000). An introduction to statistical inference – 3. *Emergency Medicine Journal*, *17*, 357–363. The examples are from a medical context but lucid and to the point.

5

INFERENCE AND REGRESSION

Chapter Overview

Introduction

Linear ordinary least squares (OLS) regression is a workhorse of observational quantitative data analysis in the social sciences. Regression takes many different forms, but its essence stays the same, so that a good grasp of the basics is fundamental. Before focusing on inference and regression, this section first introduces the analysis of variance (ANOVA) and the F distribution and then briefly reviews the components of regression to pave the way for the discussion of inference. Feel free to skip this session if you are familiar with regression. I cover only the most fundamental aspects of regression in this chapter. Consult some of the excellent guides to regression such as Allison (1999) to get a fuller picture.

The analysis of variance

We saw back in Chapter 2 that the **variance** of a variable (the square of the standard deviation) is the average of the sum of squared deviations from the mean, so that it describes how far its values are dispersed from or cluster toward the mean:

$$\sigma^2 = \frac{\Sigma(X - \mu)^2}{n}$$

When we estimate the variance using a sample, we use a denominator of $n - 1$ to get an unbiased estimate of the population variance since the amount of variance is constrained by the size of the sample:

$$s^2 = \frac{\Sigma(x - \bar{x})^2}{n - 1}$$

ANOVA was developed by Fisher as a way of examining association in a sample between a continuous variable and a categorical variable taking more than two values. This could be done by comparing the variation *within* each group of cases defined by the values of the categorical variable to the variation *between* the groups, as measured by the variance. This was preferable to making multiple comparisons between pairs of means (e.g. by using t or z tests for significance), because multiple comparisons complicate the estimation of p-values as soon as more than one hypothesis is being tested with the same data. Let our categorical variable be x and our continuous variable be y. The null hypothesis is that all the groups are samples drawn randomly from the same population of the variable y, because if this is the case, then there can be no association between y and the grouping variable x.

ANOVA proceeds by comparing the variance *between* group means defined by the categories of x to the variance of y *within* each of these groups, in each case using the sum of squared deviations (the numerator in our variance formula above) referred to as the **sum of squares**. It is the marginal or grand mean value of y and its conditional mean value for each group that is the focus of interest.

If the null hypothesis is true, then both the **between-group variance** and **within-group variance** are estimates of the population variance, so that their ratio would be equal to 1. The unconditional variance of y would be the same as the variance of y, conditional upon each of the values of x. Alternatively, if the variance *within* groups is small compared to variance *between* them, then the variance of y conditional upon the categories of x will be less than the unconditional variance, and we will have evidence of association between x and y: some or all of the variance in y will be accounted for by x. You may find it helpful to think of a limiting case in which all the variance in y is accounted for by x. There would be *no* variance within each group defined by x, the **conditional variance** of y would be 0 for each category of x. All the variance would be between the groups defined by x.

The ratio of the sums of squares for the between-group variance to the within-group variance produces the *F-statistic*, which has a given distribution for any degrees of freedom in each of the two sums of squares, just as we saw with the χ^2 statistic. The *greater* the value of F for any combination of degrees of freedom, the higher the ratio of the between-group variance to the within-group variance, the stronger the evidence of association and the smaller is the associated probability of observing that data under the null hypothesis that the means are all equal. Any statistical software will calculate the value of F from the appropriate distribution:

$$F = \frac{\text{Unconditional or between-group variance}}{\text{Conditional or within-group variance}} = \frac{\text{Variance accounted for by } x}{\text{Unaccounted variance}}$$

By far the best way to understand ANOVA is to work through an example. The National Child Development Study (NCDS) had the children take a maths test at age 11. In order to keep the numbers small and calculations clear, I've taken a subsample of boys' scores in this test, divided into three groups by the social class of their father at birth. Table 5.1 and Figure 5.1 show the main results and the calculations used to arrive at them. Figure 5.1 is a dot plot for the scores with a panel for each category of father's social class. The maths scores for all 100 boys (our y variable) have a grand mean of 18.3 and a standard deviation of 11.5. This grand mean is shown as a thick grey dotted line in Figure 5.1. We have three *conditional* means for maths scores for each category of boys defined by our x variable, father's occupational class. These are shown as solid black lines in Figure 5.1 and in the third column of Table 5.1.

Figure 5.1 illustrates the variance of the boys' scores within each social class category if we examine each panel separately, and unconditionally, for all boys, if we look at all three panels together. The black arrows indicate the absolute deviations from the group means for a selection of cases in the plot, which have been highlighted in black, while the grey arrows indicate the absolute deviation of each of the three group means from the grand mean of 18.3.

Table 5.1 NCDS boys' maths test scores at age 11 by fathers' social class

Social Class	N	M	SD	Variance	Sum of Squares (Variance * (n − -1)), Within Group	Deviation From Grand Mean	Deviation Squared	Sum of Squares (Variance * n), Between Group
Non-manual	23	24.13	11.71	137.1	3016	5.85	34.2	787
Skilled manual	58	17.47	10.92	119.2	6792	−0.81	0.66	38
Semi/unskilled manual	19	13.68	10.90	118.8	2138	−4.60	21.12	401
Total	100	18.28	11.54		11,947			1227
Mean square					11,947/97 = 123			1227/2 = 614

Note. NCDS = National Child Development Study. Data from NCDS teaching data set.

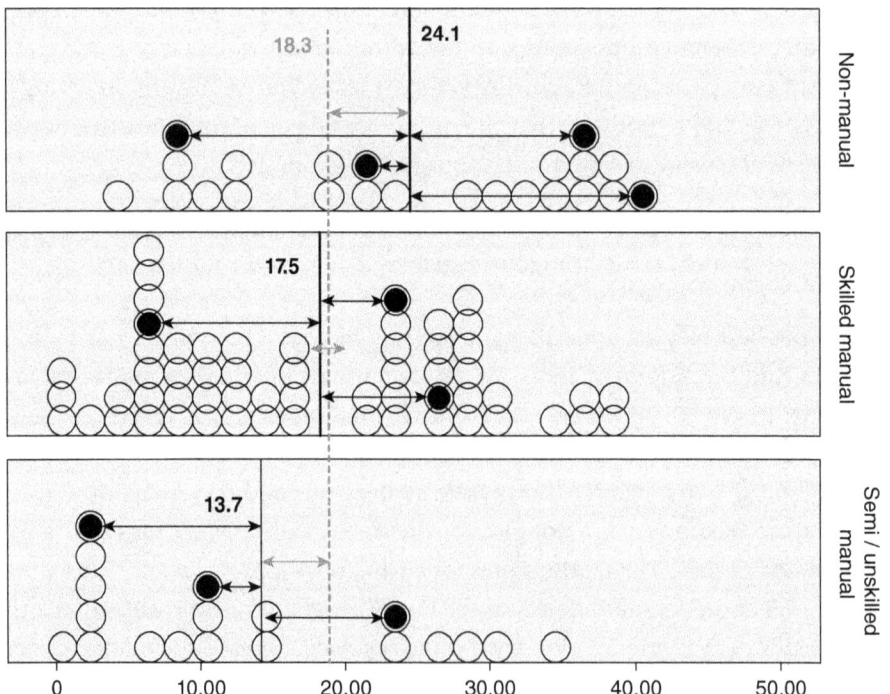

Figure 5.1 NCDS boys' maths test scores at age 11 by fathers' social class

Note. NCDS = National Child Development Study. Data from NCDS teaching data set.

Children of fathers who were non-manual workers scored 24.1 on average compared to 17.5 for those with fathers who were skilled manual workers and 13.7 for those with fathers who were semi-skilled or unskilled manual workers. But given the small numbers, it could be that these differences between the group means are due to sampling variation. Our null hypothesis is therefore that their maths scores were *not* associated with father's social class, so that these scores are drawn from a single distribution of scores. To test this null hypothesis, we need to calculate the sums of squares for the variance within our three groups and compare it to the sum of squares for all three groups together.

To calculate the *within*-group sums of squared deviations from the mean, we calculate the variance (column 5) for each group by squaring the standard deviation (column 4) and then multiplying by $n - 1$ (from column 2) for the group to remove the denominator from the variance calculation (because both the standard deviation and the variance will have been calculated using the formula for samples in which the sum of squared deviations is divided by $n - 1$). The result for the *within* sum of squares for each group is shown in column 6 of Table 5.1. These values correspond to the square of the residuals from the conditional mean for each group of cases, represented by the black arrows in each panel of Figure 5.1. We then divide this total within-group sum of squares by the *degrees of freedom*. We have 100 observations divided into three groups, so that we would need to define 97 values to leave the others fixed; our choice for the final three scores would be constrained by the need to keep the group and grand means. This gives us the **mean square within groups** of 123 (Table 5.1, bottom row, column 6). This gives us the denominator for the calculation of our *F*-statistic: variance conditional upon and therefore unaccounted for by fathers' class.

To calculate the sum of squares *between* groups, we take the deviation of each group mean from the grand mean for all groups (shown in column 7 of Table 5.1 and represented by the three grey arrows in Figure 5.1), square it (column 8) and multiply by the number of observations (column 2). As we did before, we total this for the three groups to produce the sum of squares and then divide by the degrees of freedom to produce the **mean square between groups**. (There are only 2 *df* since the third sum of squares will be defined by the total minus any of the two sums of squares that have been defined.) This gives us the numerator for the calculation of our *F*-statistic: the unconditional variance of *y*.

The ratio of the between-group variance to the within-group variance thus comes out as 5.

$$F_{2,97} = \frac{614}{123} = 5.0$$

Figure 5.2 shows the *F* distribution for these degrees of freedom. The portion of the line marked in bold shows cumulative probability of a value of *F* as high as 5 under the null. It is negligible ($p = 0.009$).

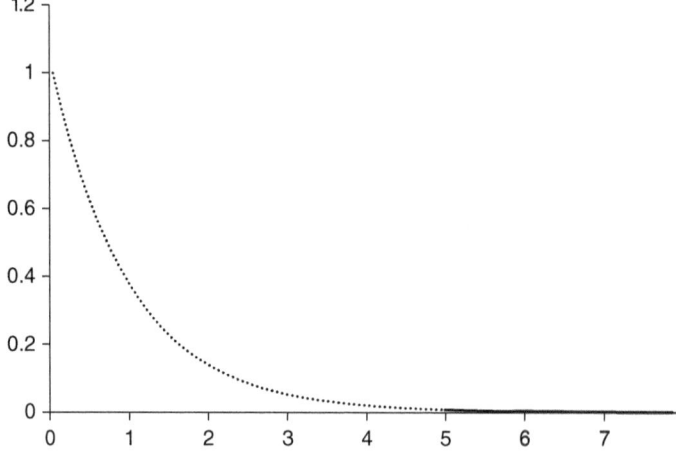

Figure 5.2 Distribution of $F_{2,97}$

The principle of ANOVA has a wider application to the association between more than one variable and a response variable. Thus, we could imagine comparing the means of a response variable broken down by more than one other categorical variable. We can ask how much of the variance of the response variable is accounted for by its association with these other variables. We can also extend the principle to association with continuous as well as categorical variables. This is what regression does. First, we will look at the concept of regression, then we see how it is a generalisation of ANOVA for categorical variables.

The regression equation

Imagine a car which runs at a constant speed of 50 km per hour, consuming a litre of petrol every 25 km. We measure how far it travels and the fuel used, and we get Table 5.2.

Table 5.2 Distance and fuel consumption

Distance Travelled (km)	Fuel Consumed (L)
5	0.2
10	0.4
15	0.6
20	0.8
25	1.0

We could represent this as a scatter plot (Figure 5.3), plotting distance on the horizontal (X) axis and fuel on the vertical (Y) axis. The coordinates would form a straight line.

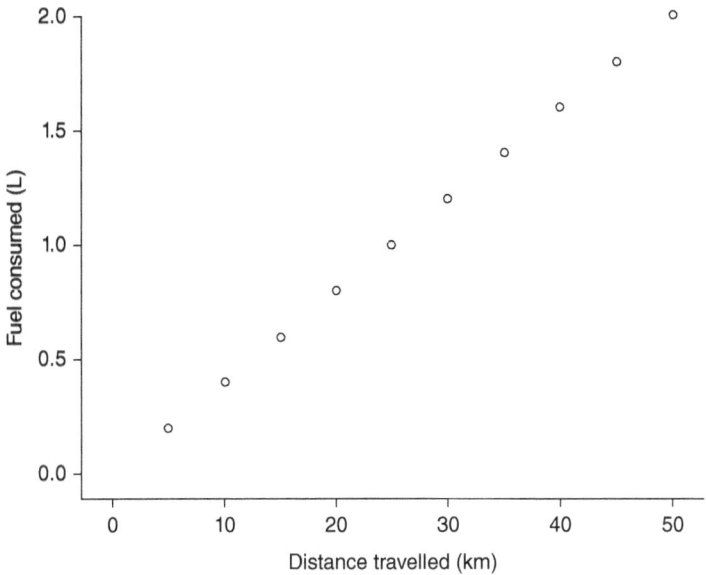

Figure 5.3 Distance and fuel consumption

Like any possible straight line we could draw on the graph, we could describe this line using only *two* numbers. The first would be the point on the scatter plot where the line crosses the vertical (*Y*) axis, when *X* is equal to zero on the horizontal (*X*) axis. This value is called the **intercept**. In our example here, when *X* is zero, the value for *Y* is also zero, so the line crosses the *Y*-axis at zero. The second number would describe the **slope** of the line: whether it rises or falls as we travel from left to right. We can calculate this second number by figuring the number of units along the horizontal *X*-axis we need to go, in order to *go up 1 unit* in the vertical *Y*-axis. In this example, we need to go 25 units along the *X*-axis to do this (which is just another way of saying that travelling 25 km consumes one litre of petrol). If we have to go 25 units of *X* to get a 1-unit rise in *Y*, this is the same as saying that if we *divide* the value of *X* by 25, we'll get the corresponding number of *Y* units. This gives our second number: 1/25 or 0.04. We can now describe our line as

$$Y = 0 + X/25 \quad \text{or} \quad Y = 0 + 0.04X$$

We've used simple algebra to express the form of the relationship between *X* and *Y*. The relationship is fixed or determined in this example, so that *Y* is a straight-forward function of *X*, because fuel consumption is entirely determined by distance travelled. We can therefore predict the value of *Y* exactly if we know the value of *X*. The intercept describes the location of our line on the scatter plot – how far up or down it lies, while our second number describes the slope. The larger the number, the steeper the slope; if the number is positive, the line slopes upwards from left to right, if it is negative, it slopes downwards. We therefore have a general equation of the form:

$$\hat{y} = \alpha + \beta x$$

where α ('alpha') is the y intercept, β ('beta') is the slope of the line and $\hat{y}$ (pronounced 'y-hat') is our predicted value for y, given our observed value for x. The line is a *regression* line and any such straight line can be described by these two numbers: the intercept and the slope.

Regression as the analysis of unconditional and conditional variance

We can use exactly the same set-up to describe the association between any two continuous *variables* that we could represent by plotting their values on each axis of a scatter plot. If our data is from a sample of any kind, then α and β would be two population parameters that we would be trying to estimate from the data in our sample. We can write the equation for the line as

$$\hat{y} = a + bx$$

The formula for b will be:

$$b = \frac{\Sigma(x - \bar{x})(y - \bar{y})}{\Sigma(x - \bar{x})^2}$$

This formula looks more intimidating than it is. The numerator comes from the formula for the **covariance** of two variables that we saw when calculating Pearson's r in Chapter 2, Section 'Pearson's r'. If values above the mean on the x variable tend to go together with values above the mean on the y variable, this will give a positive sum, both from positive and negative residuals. If values above the mean on x tend to go with values below the mean on y, we have a negative amount, so that the sign for b will be positive for an upward sloping line and negative for a downward one. The magnitude of b will come from the relative distance to the mean of the corresponding values of x and y; in other words, the average amount of difference in y associated with a change in the value of x, as we saw before. As usual, there is no need to memorise this equation, but it is one we will return to a little later.

Figure 5.4 shows a scatter plot of cell phone ownership (contracts per 100 population) on the horizontal (X) axis by mean life expectancy at birth (E_0) plotted on the vertical (Y) axis for the countries of the world in 2017, using World Bank data from the *Gapminder* website (www.gapminder.org). Each coordinate on the plot represents the combination of values for these two variables for one country. Clearly, the coordinates do not lie on a single straight line. It would be very odd if they did so, as we would not expect there to be any kind of deterministic link between the level of life expectancy in a country and the volume of cell phone subscriptions.

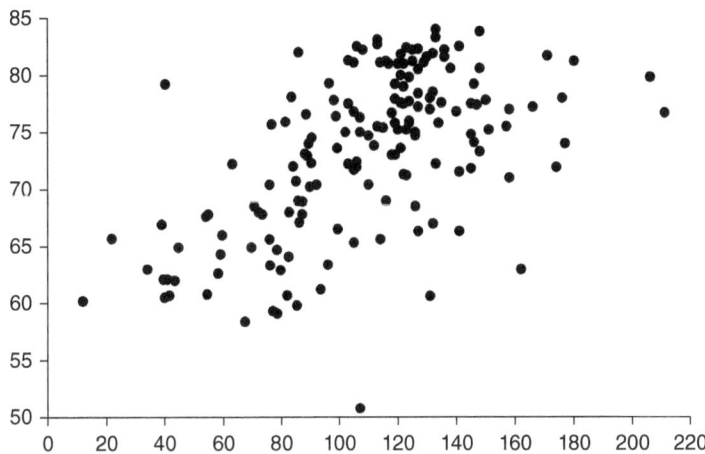

Figure 5.4 Cell phone subscriptions and mean life expectancy at birth for countries in 2017

Note. www.gapminder.org

However, it is also clear that there is a pattern to the coordinates. In countries where there are more cell phone subscriptions, life expectancy tends to be higher. There are almost no countries in the top left quadrant with few cell phones and high life expectancy (the single exception is Cuba), while there are also few countries in the bottom right quadrant. We could therefore still summarise the association with a straight line, but we need a way of defining where it should go.

A good way would be to put the line where it *minimised* the sum of the squared vertical (*y*) distances to the *coordinates*, analogous to the way in which the unconditional variance of *y* uses the squared vertical distances to $\bar{y}$. This is the ordinary *least* squares **regression line**. It is the line which minimises the sum of squared vertical distances (residuals) between that line and all the coordinates. Figure 5.5 illustrates this idea (for a selection of about one third of the coordinates, so that it is easier to see). Each vertical dotted line represents the residual: the distance between the *observed* value for *y* and the value for *y* we would *expect* if it lay on the regression line minimising this *sum of squares*. What is this expected value? If our line summarises the position of the coordinates, it must represent the value we would expect for *y*, conditional upon the value we have for *x*, or in other words, our *estimate* for *y*, once we know and can take account of the value of *x*. This regression line gives us estimates for the value of y based on the value taken by x, represented by $\hat{y}$ (pronounced *y-hat*). Our *sum of squares* has the formula:

$$\Sigma(y-\hat{y})^2$$

Recall from Chapter 2 that the symbol Σ means 'take the sum of' all the terms to its right.

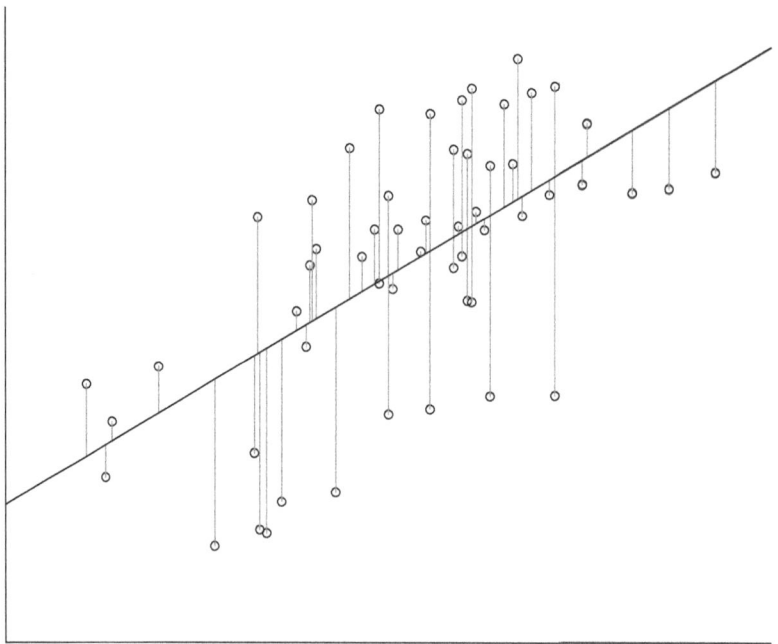

Figure 5.5 Residuals from coordinates to the regression line

As before, we can calculate the slope of the regression line by working out how far along the X-axis we have to go to obtain a 1-unit increase in the Y-axis. As before, the point where our regression line crosses the vertical axis (the value of $\hat{y}$ when x is 0) will be our intercept. The regression equation for Figure 5.4 turns out to be:

Life expectancy at birth (years) $= 61 + 0.113 *$ Mobile subscriptions

It predicts that a country with 150 subscriptions per 100 population would have a mean life expectancy at birth of $61 + 0.113 * 150 = 61 + 17 = 78$ years, where 61 is our estimate of the α parameter, while 0.113 is our estimate of β, the *regression coefficient*. Our example illustrates the danger of using regression coefficients to make causal arguments. We have clear evidence of an association between life expectancy and the diffusion of mobile phone, but a causal link in either direction would be highly improbable. Both are likely to be a function of a third (unmeasured) variable: level of economic development.

The $\hat{y}$ value at any point on the regression line represents the mean values of y, *conditional upon* the corresponding value of x. For any $\hat{y}$ value, there will also be a conditional standard deviation, describing the spread of the y coordinates around $\hat{y}$ for that value of x, that is, how closely the coordinates cluster near the line or disperse far from it. To examine this, let's look at some NCDS data.

Figure 5.6 shows the regression of the heights of the women in the study at age 23 (in centimetres) on the height of their fathers (in inches) when they were born (since heights were measured to the nearest inch more than one coordinate may fall on the same point within the plot). The coordinates for the height of the children whose fathers were 72 inches (6 feet) tall have been highlighted in solid black.

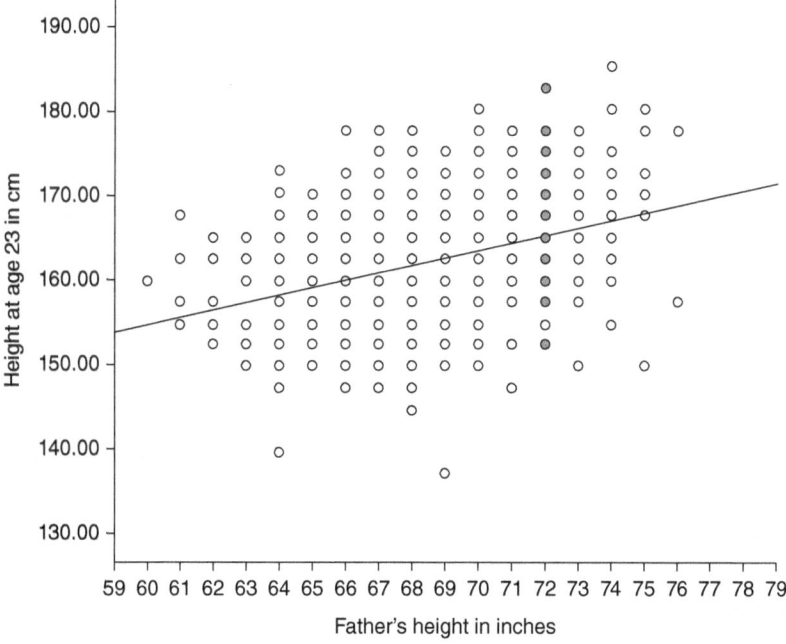

Figure 5.6   Daughter's height at age 23 (in cm) by father's height at birth (inches)

Note. National Child Development Study teaching data set.

Seventy-nine girls had fathers of this height. Their mean height (in cm) at age 23 was 166.2 cm ($\hat{y}$) when $x = 72$. The standard deviation of their heights was 6.5 cm (calculated using the squared residuals from their mean height as we've seen already). This is the *conditional* standard deviation of y. The histogram in Figure 5.7 shows the distribution of the heights of these girls. The OLS model assumes that the observed values of y conditional on each value of x are distributed normally and that their standard deviation is the same. That is to say that the spread of values for y is similar at different values of x. In practice, however, the estimates for regression coefficients tend to be robust to violations of these assumptions.

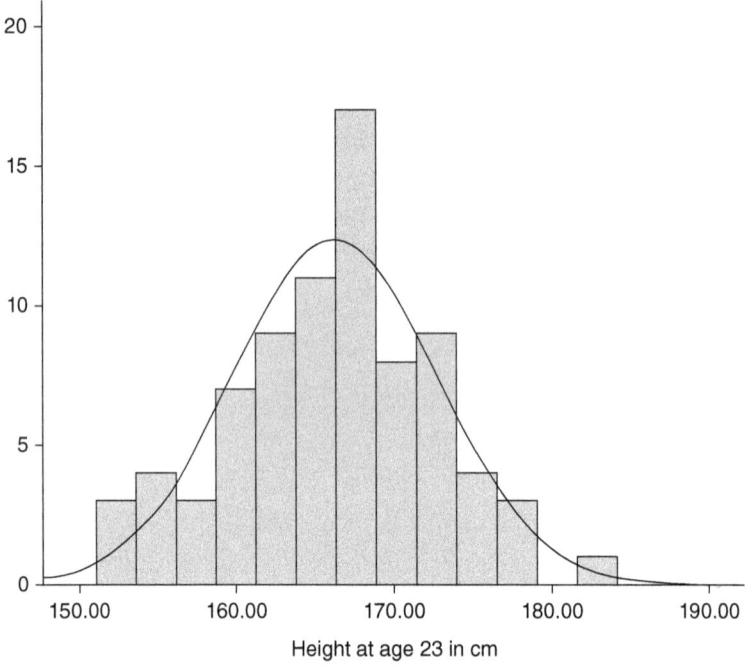

Figure 5.7 Fathers 6 feet tall: daughter's height (in cm) at age 23

Note. National Child Development Study teaching data set.

Once we have estimated a regression equation and its corresponding line, we have a *third* quantity that we can evaluate: how closely do our coordinates cluster around our regression line? This is the same as asking the question 'how large is the variance of y *conditional upon* x, compared to its *unconditional* variance?' This is the question we have already looked at when we calculated the F-statistic in ANOVA in Section 'Analysis of Variance'.

We saw one 'slice' of conditional variance in the histogram of girls' heights, conditional upon their fathers' height in Figure 5.6. You could imagine a succession of these histograms placed on their sides, centred on the regression line, for each separate inch of fathers' heights. Imagine our two limiting cases. If there was a perfect association between father's height at birth and daughter's height at age 23, the conditional variation of daughter's height would be zero, since all the daughters of fathers of any given height would have the same height, and all the coordinates in our scatter plot would lie precisely on a straight line. Conversely, if there was no association at all between father's height and daughter's height at age 23, so that our coordinates were randomly scattered about the graph, the variation of daughters' heights and the variation, conditional upon the height of father, would be the same.

You may also find the analogy with a contingency table for two discrete variables helpful. When there is no association between two categorical variables x and y, the marginal distribution of y is the same as its distribution conditional upon any value of x. If you are unsure about this point, you may like to review Chapter 2, Section 'Conditional Probabilities With Categorical Variables'.

The ratio of the *conditional* variance of y *compared to* its **unconditional or marginal variance** gives us a measure of the association between the two variables in terms of a **correlation coefficient**, Pearson's r, and a measure of the proportion of the variance in y accounted for by the variance in x known as the **coefficient of determination or R^2**.

Correlation and regression

A useful way to think about this is that *correlation* describes the clustering of the coordinates around the regression line, while regression coefficients describe the position of the line: its *intercept* and *slope*. If the correlation is high, points will cluster close to the line and the conditional variance will be small compared to the marginal variance. Conversely, if there is little correlation between x and y, the conditional variance will approach the marginal variance. Were there to be zero correlation, the conditional and unconditional variance would be exactly the same.

In order to calculate the coefficient of determination, we take the sum of squares for the residuals from the coordinates to the regression line and divide it by the sum of squares for the residuals from the coordinates to the mean of y. This gives us a value from 1 (when there is zero correlation between x and y) to 0 (where there is perfect correlation between x and y such that knowing the value of x perfectly predicts the value of y). It makes sense to reverse this order, so that 0 corresponds to no correlation and 1 to perfect correlation. We do this by subtracting the fraction we have just calculated from 1.

We have the sum of squared errors or *residual sum of squares* from the *conditional* variance:

$$\Sigma(y - \hat{y})^2$$

[take each residual from the regression line $(y - \hat{y})$, square it and then add them all up]

We have the *total sum of squares* given by the residuals from observed values of y to its mean, or the *unconditional* variance:

$$\Sigma(y-\bar{y})^2$$

[take each residual from the mean $(y-\bar{y})$, square it and then add them all up]

So our formula for R^2 becomes:

$$1-\left(\frac{\Sigma(y-\hat{y})^2}{\Sigma(y-\bar{y})^2}\right)$$

As with ANOVA, you may find it helpful to imagine two limiting cases. One would be where all the coordinates fell on the regression line (like our example of the car consuming petrol). Knowing the value of x would allow us to fix the value of y with absolute certainty. The variance of y, conditional on x, would be zero.

Alternatively, we can imagine what would happen if there was absolutely *no association* between x and y. Knowing the value of x would give us absolutely no information about the value of y. How might we best predict the value of y for any value of x in such a situation? Our best bet would simply be to guess the mean value of y, $(\bar{y})$ regardless of the value of x. Our regression line in such a situation would be a horizontal line crossing the Y-axis at the mean value of y.

If we take our *residual sum of squares* as a proportion of the *total sum of squares*, we have a formula for expressing the proportion of the variance in y that is *not* accounted for by the variance in x. Test this out with our limiting cases. If all the variance in y was accounted for by x, that is, if all the coordinates lay on the regression line, then the residual sum of squares would be 0 and the value of our formula would be 1:

$$1-\left(\frac{\Sigma(y-\hat{y})^2}{\Sigma(y-\bar{y})^2}\right)=1-\left(\frac{0}{\Sigma(y-\bar{y})^2}\right)=1-0=1$$

Conversely, if there was no association between x and y, the regression line would simply be the mean of y, none of the variance in y would be accounted for by x, and $\hat{y}$ would exactly equal $\bar{y}$ so that our expression would be equal to 0:

$$1-\left(\frac{\Sigma(y-\hat{y})^2}{\Sigma(y-\bar{y})^2}\right)=1-\left(\frac{\Sigma(y-\bar{y})^2}{\Sigma(y-\bar{y})^2}\right)=1-1=0$$

Both r and R^2 can take any absolute value between 0 and 1, and r can take either a negative or a positive sign. You can think of the value of r as a 'standardised' slope, representing the way in which a 1 standard deviation change in the value of x is associated with a change of r standard deviations in the predicted value of y ($\hat{y}$).

Regression and ANOVA

We can see how linear regression and ANOVA use the same statistics logic and calculations by looking at our boys' maths scores as a regression. To do this, we must first introduce the concept of a **dummy variable**. The predictor or independent variables in linear regression are assumed to be at the interval level of measurement, which creates a challenge for dealing with categorical variables. We get around this difficulty by splitting up each categorical variable into a series of *dummy* variables, each of which takes the value of either 0 or 1 for each case. We can account for all the categories in a variable by creating fewer dummies than there are categories in our original variable. Table 5.3 shows how this is done with the variable for fathers' social class, splitting it into two variables *classnm* and *classsm*. Our dummies code each of the first two class categories as '1'. This leaves the third category to be identified by the cases taking a '0' for both our dummies.

Table 5.3 Dummy variables for fathers' class

Categories: Father's Social Class	n	Classnm	Classsm
Non-manual	23	1	0
Skilled manual	58	0	1
Semi-/unskilled manual	19	0	0

Note. National Child Development Study teaching data set.

Table 5.4 shows some of the output from IBM SPSS Statistics for the regression of maths scores on fathers' class for our 100 boys. Most of the output should look familiar from the calculations set out in Table 5.1.

Table 5.4 IBM SPSS statistics regression output

R	R Square	Std. Error of the Estimate
.305[a]	.093	11.09804

	Sum of Squares	df	Mean Square	F	Sig.
Regression	1227.015	2	613.508	4.981	.009[b]
Residual	11947.145	97	123.166		
Total	13174.160	99			

	Unstandardized Coefficients		Standardized Coefficients		
	B	Std. Error	Beta	t	Sig.
(Constant)	13.684	2.546		5.375	.000
classsm	3.781	2.934	.163	1.289	.200
classnm	10.446	3.441	.383	3.036	.003

The value for R^2 is given by:

$$1 - \frac{\text{Residual sum of squares}}{\text{Total sum of squares}} = 1 - \frac{11947}{13174} = 0.093$$

SPSS uses the terms *regression* and *residual* mean square to refer to the mean square *between* and *within* groups. The value for the *F*-statistic is given by:

$$\frac{\text{Regression mean square}}{\text{Residual mean square}} = \frac{614}{123} = 5.0$$

The constant or intercept gives us the mean maths scores for those cases coded as '0' in each of our dummies – that is, sons of semi or unskilled manual workers: 13.68. On an average, being the son of a skilled manual worker (Classsm) raises the average score to 13.68 + 3.78 = 17.47 (after rounding) while being the son of a non-manual worker (Classnm) raises the mean score to 13.68 + 10.45 = 14.13. These are the results we saw in Table 5.1. However, our regression equation also gives us separate standard errors for our regression coefficients. As usual, we can convert these standard errors to *t* scores (= B/SE) and thus a probability for observing such a value in our sample were the real coefficient in the population equal to zero. We have good evidence here that being the son of a non-manual worker did have an impact on performance in the maths test. However, we would not be safe in drawing the same conclusion for sons of skilled manual workers.

Multiple regression

Although we can only visualise two variables on a two-dimensional scatter plot, the mathematics of regression work for *any* number of predictor variables, so that the regression equation becomes

$$\hat{y} = a + b_1x + b_2x + b_3x + \ldots$$

The intercept is still the value of $\hat{y}$ when the value of all independent variables equal zero. The *regression coefficients* b_1, b_2, b_3 ... now express the change in $\hat{y}$ associated with a 1-unit change in that predictor variable, *holding constant, or controlling for,* all the other predictor variables. This is the feature that gives regression its power.

Consider a regression equation describing graduate earnings. It is likely that among the independent variables accounting for earnings would be factors such as occupation, the subject of degree and gender. It would be impossible to disentangle these factors by looking at average earnings broken down by each of them in turn.

The comparison of mean earnings of engineers with those of doctors would be contaminated by the fact that the majority of doctors are women, while most engineers are men. The comparison of men to women would be contaminated by the fact that the distribution of men and women across degree subjects differs substantially. Comparisons across degree subjects would be contaminated by the fact that graduates in some subjects tend to go into a fairly narrow range of jobs. Few medicine graduates work as engineers, and thankfully, few engineers become doctors. It would be possible to laboriously work out the average earnings of every possible combination of our independents, such as male medical graduates working as engineers, but we would then find that the numbers in many of our combinations would be so small that random noise would swamp any signal. So long as we expect our associations to be linear in form, regression gets around this by enabling us to estimate the average effect of each of our independents, net of all the others. This is its power.

The regression coefficients for each independent variable are expressed in the units of that variable, so that comparisons between them are difficult to make. **Beta coefficients** re-express the original coefficients in units of standard deviation, giving the change in standard deviation units of $\hat{y}$ for a 1 standard deviation change in b_1, controlling for the values of b_2, b_3, ... The coefficient of determination R^2 will be the square of the correlation between y and $\hat{y}$ and will express the proportion of the variance in y accounted for by all the predictor variables together.

The most important word in linear ordinary least squares regression is *linear*. If a straight line is *not* a good summary of the relationship then the estimates that we will get for the correlation between our variables and for any regression coefficients will be wrong. That is one reason why the first stage in any regression is to produce a scatter plot of each of the x variables with the y variable to check for linearity. This also allows you to screen for outliers: data coordinates that lie far from others in the plot. There are three other assumptions that statistical inference requires. The first, as always, is that the sample data is either a random sample of the target population or a convincing approximation to it. The second and third are that the conditional distribution of y for each value of x is approximately normal and that the standard deviation is the same. In practice, OLS regression is usually robust to violations of these final two assumptions.

Inference in regression: the *F*-statistic and standard errors

Recall that when we used the χ^2 distribution to infer association in contingency tables from sample data, we used the table marginals to calculate the number of observations we would *expect* to see in each table cell under the null hypothesis condition of no association. If what we observed was sufficiently different from that, so that our observed data was sufficiently improbable under the null, then we accepted we had evidence of association. We could also use confidence intervals to express the likely margin of error around any estimate of a parameter. Similarly, in OLS regression, we establish a null hypothesis, which is that $b = r = 0$ and use it to calculate the probability of observing our data were that hypothesis true. If we discard the hypothesis, because that probability is low enough, we have provisional evidence of something else going on. We also use standard errors to place confidence intervals around our estimates of the intercept (a) and our regression coefficients (b_1, b_2, b_3, etc.).

Recall from Section 'Regression as the Analysis of Unconditional and Conditional Variance' that the coefficient of determination (R^2) is given by the ratio

$$1 - \left(\frac{\Sigma(y - \hat{y})^2}{\Sigma(y - \bar{y})^2} \right) = 1 - \frac{\text{Residual } SS}{\text{Total } SS}$$

We can re-arrange this as

$$R^2 = \frac{\text{Total } SS - \text{Residual } SS}{\text{Total } SS} = \frac{\text{Regression } SS}{\text{Total } SS}$$

SPSS output also reports the **regression sum of squares**, which is the **residual sum of squares** subtracted from the **total sum of squares** or the numerator of our second equation above. The numerator ($\Sigma(y - \hat{y})^2$) captures how tightly the coordinates cluster around the regression line, while the denominator $\Sigma(y - \bar{y})^2$ captures the total variance in the response variable. The tighter the clustering, the smaller the value of the numerator becomes; the greater the variance to be explained, the larger the value of the denominator, so that the better the regression line summarises the association between the predictor and response variables, the smaller the value of the fraction becomes, and the values of R^2 and r move closer to 1.

As we saw when we looked at ANOVA, the F-statistic is the ratio of two variances, that of the unconditional variance in y and the variance in y conditional upon the predictor variables. Like the χ^2 statistic, the shape of the sampling distribution of F depends upon degrees of freedom terms. The first degree of freedom term is equal to the number of predictor variables in the model (k) and the second $(n - (k + 1))$ is equal to the sample size (n) minus the number of parameters in the model. Because there is the α (intercept) term in the model, there is one more parameter than the number of predictor variables. The larger the value of R^2, and the larger the sample size, the higher will be the value of F. As with other test statistics, the associated p-value gives the proportional area of the distribution of the statistic that comprises the critical region to the right of the value of the test statistic, which gives the probability of observing the results in our data, conditional upon the null hypothesis of $b = r = 0$ being true.

The standard error of the unstandardised regression coefficient b is given by the formula:

$$\frac{\sqrt{\frac{\Sigma(y - \hat{y})^2}{n - 2}}}{\sqrt{\Sigma(x - \bar{x})^2}}$$

We use $n - 2$ in this equation as we have lost 2 df (1 for the mean and 1 for the sum of squares). Examining this formula shows that standard error gets smaller as:

- the conditional variance of y *reduces* (i.e. as changes in the value of x account for more of the variation in y),
- the square root of the sample size *increases* (as we have already seen, to increase the accuracy we expect from a sample, we need to increase its size by the *square* of that increase: to triple the accuracy, we need a sample $3^2 = 9$ times larger), and
- the variance of x *increases* (so that there is a wider spread of values from which to predict the value of y).

Recall from Chapter 3 that the standard error gives us an estimate of the impact of sampling variation, so that dividing our estimate for b by the standard error will give us a test statistic and an associated p-value. In regression software, the t-distribution is always used (since a regression might be run on any sample size), but remember that for samples above about $n = 30$, t and z rapidly converge.

Confidence intervals for standardised regression coefficients or r are slightly more complicated by virtue of the fact that the range of absolute values of r or *beta* is constrained to lie between −1 and +1. This is not a transformation or calculation that you will ever have to carry out by hand; you can rely on software to calculate for you.

An example of multiple OLS regression

As usual, the best way to understand all this is to work through an example. We have information on the gross weekly earnings for 404 women in the NCDS study who were working at age 23. Table 5.5 describes this variable and also four predictor variables that we might expect to be associated with earnings:

1   School type dummy variable: 0 = secondary modern or comprehensive; 1 = grammar or independent school
2   Father's class dummy variable: 0 = manual; 1 = non-manual
3   Maths test score at age 16 (score out of 40)
4   Height at age 23 in cm

Table 5.5 Regression variables

		Mean	SD	N
y	Gross weekly earnings	81.56	27.27	404
x_1	School dummy	0.228	0.420	404
x_2	Father's class dummy	0.280	0.4490	404
x_3	Maths test score at age 16	13.47	6.91	404
x_4	Height at age 23 in cm	163.3	6.87	404

The scatter plots of the predictor variables by our response variable shown in Figures 5.8 and 5.9 suggest that the association between the continuous predictors and the response variable is approximately linear, or not obviously non-linear. Tables 5.6 to 5.8 report the results of running a multiple linear regression model. Depending on the statistical software you use, the values reported here may appear in a different order or be labelled slightly differently. These tables have been drawn from IBM SPSS output.

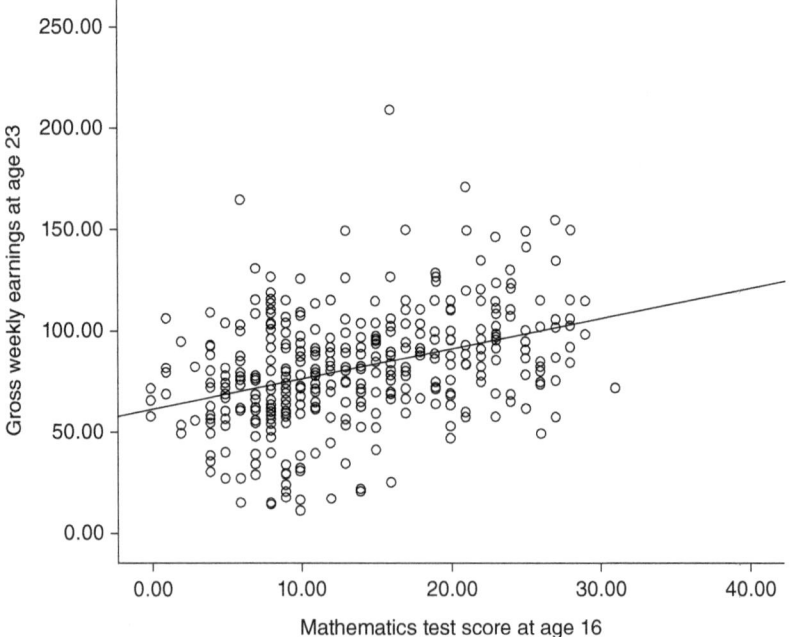

Figure 5.8 Earnings by maths test score

Note. National Child Development Study teaching data set.

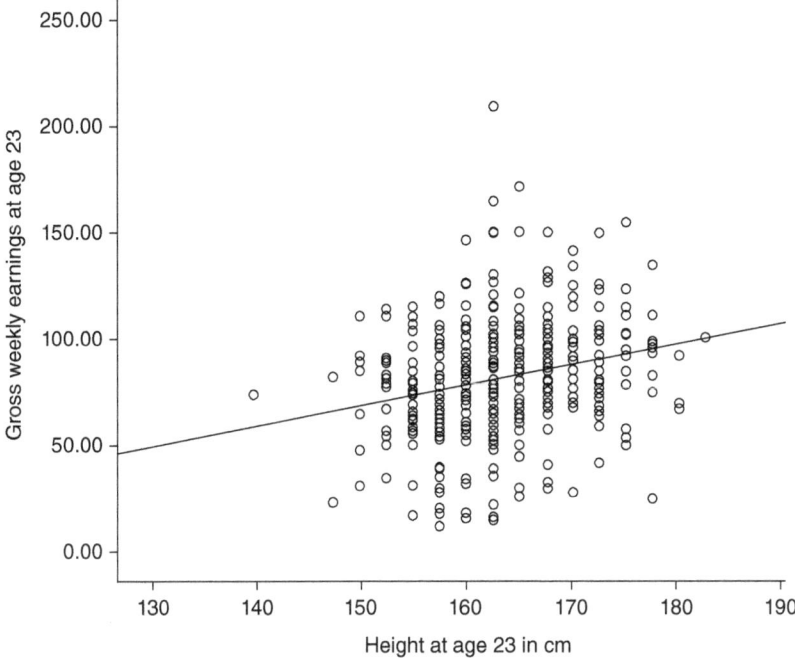

Figure 5.9   Earnings by height in cm

Note. National Child Development Study teaching data set.

Table 5.6   Correlation and coefficient of determination

R	R Square	Adjusted R Square	Std. Error of the Estimate
.437[b]	.191	.183	24.65411

Table 5.7   Analysis of variance and sums of squares

	Sum of Squares	df	Mean Square	F	Sig.
Regression	57272.586	4	14318.146	23.556	.000[c]
Residual	242522.214	399	607.825		
Total	299794.800	403			

Table 5.8 Regression coefficients and associated p-values

| | Unstandardised Coefficients | | Standardised Coefficients | | |
	B	SE	β	t	Significance
(Constant) Intercept	−33.171	29.726		−1.116	.265
School dummy	8.503	3.369	.131	2.524	.012
Class dummy	4.574	2.921	.075	1.566	.118
Maths test score at age16	1.037	0.206	.263	5.038	.000
Height at age 23 in cm	0.597	0.185	.150	3.236	.001

Table 5.6 gives us the correlation r (0.437) between y, the observed values for gross earnings, and $\hat{y}$, the values for gross earnings predicted by our regression model. The square of r, R^2, is also equal to the regression sum of squares (57,272) divided by the total sum of squares (299,794) shown in Table 5.7. Under 'Standard error of the estimate', SPSS reports the root mean square, which is equal to the standard deviation of gross earnings *conditional upon* the predictor variables. If you divide the standard deviation of $\hat{y}$ by the standard deviation of y, you will get the value for r. You can obtain the 'Standard error of the estimate' by the formula

$$S = \sqrt{\frac{\text{Residual } SS}{n-(k+1)}} = \sqrt{\frac{242522.214}{(404-5)}} = 24.65$$

The value for adjusted R^2 takes account of the number of variables in the model. Under the 'Mean square' column in Table 5.7, we have the relevant sums of squares divided by the degrees of freedom, and their ratio (14,318/608) produces the F-statistic with its corresponding p-value. This tells us that at least one of the regression coefficients in the model is *not* equal to 0. In other words, our model improves on just taking the mean of y as our estimate, without the predictor variables.

The value for the intercept (−£33.17p) reported in Table 5.8 tells us our estimate ($\hat{y}$) of average earnings for women who went to a secondary modern or comprehensive school, whose father's occupational social class at the time of their birth was manual, who scored 0 in the maths test at age 16 and had a height of 0 cm. Nobody is 0 cm tall and nobody earns 0 wages. Regression estimates are only viable for the range of data values that we have observations for. This often makes the raw intercept of limited interest when there are continuous predictor variables in the equation. Substituting a height of 163 cm and a mean maths score of 13, we'd get an estimate for average earnings of

$$-£33.17 + (1.037 * 13) + (0.597 * 163) = -£33.17 + (13.48) + (97.31) = £77.62$$

Having gone to a grammar or independent school would increase that estimate by £8.50 and having a non-manual class father by £4.57.

The absolute size of the regression coefficients in Table 5.8 is a function of their units of measurement. For both dummy variables, the absolute value of the unstandardised coefficient gives us the estimated change in the mean gross earnings associated with being in each category of the dummy variable. By contrast, the coefficient for height is for a 1-cm change in height. The standardised coefficients adjust for this by re-expressing the regression coefficient in terms of standard deviations. The larger the standardised regression coefficient, the bigger the impact of the predictor. The standard deviation of women's earnings was £27.27. The standard deviation of maths score was 6.9. Thus, a 6.9 (= 1 SD) increase in maths score should lead to an increase in average earnings of 0.263 * 27.27= £7.17. You can check this by doing the calculation using the regression coefficients: 6.9 * 1.04 = £7.17.

These four variables account for about 18% of the variance in earnings. However, these estimates are for our data. We also need to think about what they mean for the target population we are generalising to. The p-values (shown under 'sig.') for each regression coefficient are produced from the t statistic, obtained by dividing the regression coefficient by its standard error. Therefore, the t statistic for the class dummy variable is 4.574/2.921 = 1.566. We do not have good reasons to reject the null for this predictor variable. We can estimate the range of likely values for its regression coefficient in the population by taking a CI of 1.96 standard errors either side of our point estimate. That gives us a range of £–1.16 to £–10.32. While there is some evidence there that the effect is positive, it is rather weak. We have stronger evidence for the impact of height, with a point estimate of about 60p gain in earnings for every extra centimetre of height. This might appear counter–intuitive, but it is a finding established in many studies. Partly this is because height is associated with class and with nutrition at early ages, and partly because height continues to be associated with status and prestige. However, although the effect is clearly significant, it is not very substantial.

What can go wrong with multiple regression?

Remember that the statistical component of inference is only one part of scientific inference. In assessing our model, we would need to consider the following:

1   *Omitted variables:* Any predictor variable associated with the response variable that is not included in the model will bias the regression coefficients. Data sources rarely contain every variable we would like to include, or measured in the way we would ideally like. Against this, too many predictors make the model clumsy to interpret

and introduce the risk of modelling noise rather than signal in the data. However, it is always good to reflect on the issue of what other variables we might expect to vary with the predictors, potentially rendering part of their explanatory power spurious.

2   *Measurement error:* Was father's social class defined consistently and accurately? Were respondents' reports of their weekly earning accurate? Did they all appreciate the distinction between gross and net earnings? What about respondents whose earnings fluctuated from week to week? We may have accurate information on school type, but does this capture what we are trying to measure? Not all comprehensives are the same!

3   *Sampling error:* The original sampling frame for the project may have allowed a rigorously random sample to be chosen, but even in such a best case scenario, missing responses and drop outs from a longitudinal study may all bias the sample. Insofar as we can treat the 1958 NCDS cohort as a random sample, we could generalise results to other female children born that year. We might also, much more cautiously, generalise to all fathers and daughters in the UK from around this time period, and more cautiously still, to all fathers and daughters. There may be many reasons why the NCDS sample differs from a perfectly random sample, and it would be necessary to consult the data documentation to get a full picture of how the study participants were recruited. It might be that selection effects arose from difference between parents who gave or withheld consent to participate in the study, or the efficiency of the study team in different areas of the country and so on. One might expect factors such as genetics, nutrition, public health, sanitation and the standard of medical care available to influence the growth of daughters and the relative impact of the genetic inheritance from fathers, so that a relationship we find in the UK in the 1950s is different from what we might discover in, for example, China in the 21st century.

4   *Linearity:* The relationship between the response variable and one or more of the predictors may not be linear. Often, some kind of transformation will address this problem.

5   *Power:* Power is discussed in greater detail in Chapter 6, but its essence is that if our sample size is too small, we will be unlikely to detect weaker associations that may be of substantive interest. Our sample size is about 400. This is reasonable, but some of our subcategories will be quite small. We only have 43 women with manual class fathers who went to grammar or independent schools.

6   *Reciprocal or reverse causation:* This is unlikely to be a problem with the model run here. It would be difficult to argue that earnings at age 23 had any causal influence on fathers' social class at birth, type of school attended or height. However, there can often be situations where the direction of cause and effect in a correlation is unclear, and this will bias the coefficients.

7   *Problems with the residual diagnostics:* This takes us beyond the scope of this book, but a good model will usually have a number of desirable features, such as patternless residuals.

Transformation

Quite often in the social sciences we have data that refers to some kind of growth process or where some kind of relatively constant rate of change is at work.

This inevitably produces situations where straight lines are not a good summary of a relationship. This is because the same amount of relative change over a period of time comes to represent ever larger or smaller amounts of absolute change captured on a linear scale. Figure 5.10 uses *Gapminder* data to plot average income in dollars against average life expectancy for the countries of the world in 2018.

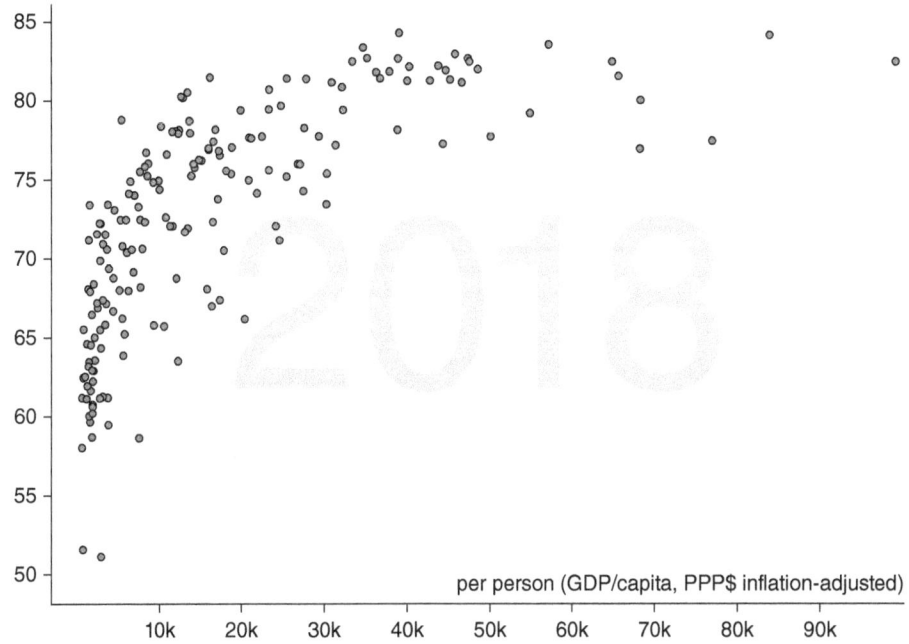

Figure 5.10   Mean life expectancy by mean income (linear scale) countries, 2018

Note. www.gapminder.org

It is clear that this is not an association between two variables that is well sum-marised by a straight line. If you think about the likely nature of the association, it would be strange if a straight line did describe it well. We could expect there to be a link between affluence and life expectancy. Poor nutrition, poor sanitation, relentless toil, crowded living conditions and few medical services to tackle dis-ease or illness are all things we might expect to depress life expectancy. However, we'd expect the impact of a change in average income from $2,000 to $4,000 to be much greater than that from $50,000 to $52,000. It is the relative change that matters, not the absolute increment. If we change our measurement of income to a log scale, where a move along the horizontal axis of our graph corresponds to a constant rate of relative change, our association straightens out nicely, as shown in Figure 5.11.

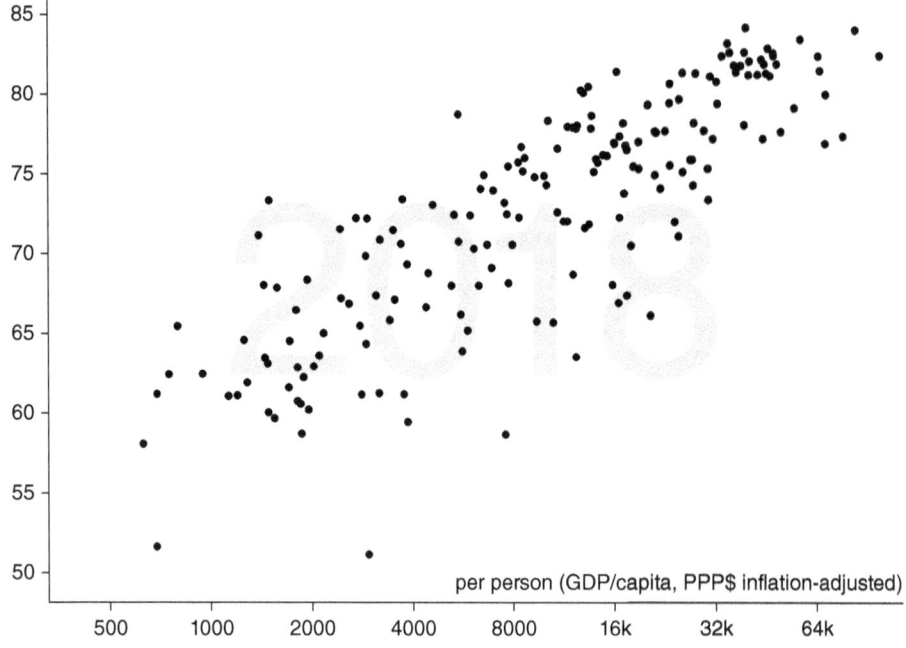

Figure 5.11 Mean life expectancy by mean income (log scale) countries, 2018

Note. www.gapminder.org

Sometimes we need to transform both variables in this way. Figure 5.12 using data from the *Gapminder* website shows the linear relationship between mean income and the metric tonnes of carbon dioxide emissions per person in that country for 2014. No relationship is apparent. However, if we use a log scale, as in Figure 5.13, it can be seen that once the variables have been transformed, there is a very strong linear association. **Logarithms** are the expression of a number to a power or exponent. They have the property that adding the logarithms of two numbers has the same effect as multiplying the original numbers. That is why logarithmic transformations of variables describing economic categories such as income are often used in regression. The appendix to this chapter explains logarithms and powers.

Appendix: A refresher on powers and logarithms

Powers

If you are unsure about logs, watch a short film *Powers of Ten*, available at www.youtube.com/watch?v=0fKBhvDjuy0.

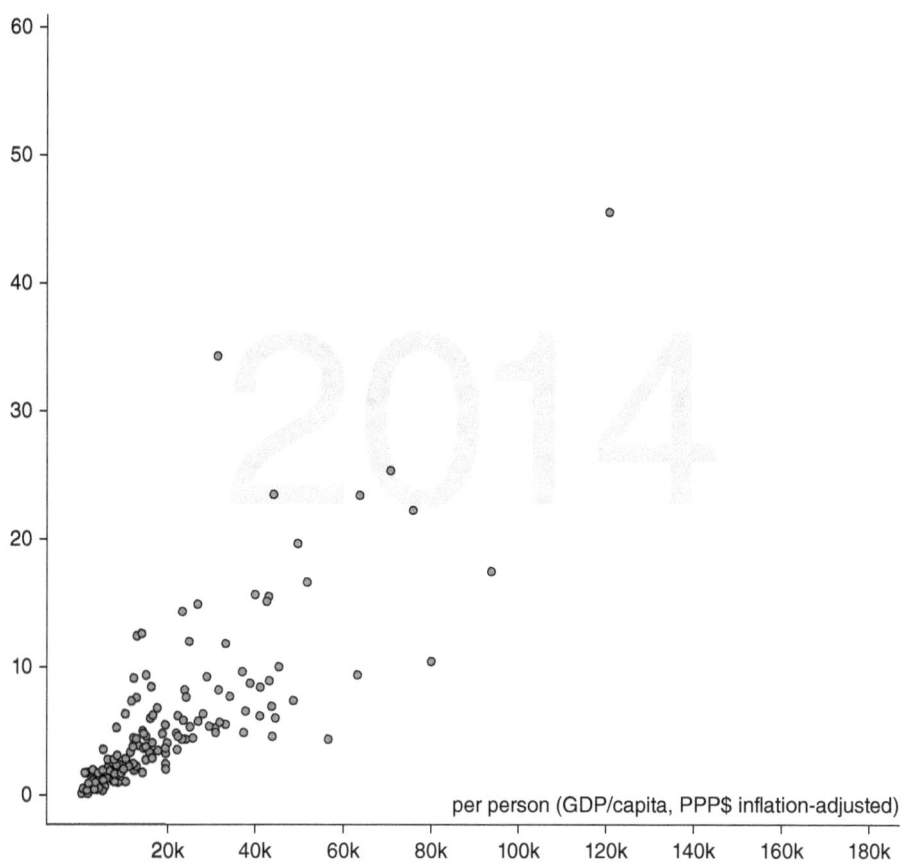

Figure 5.12 Mean income and carbon dioxide emissions (linear scale)

Note. www.gapminder.org

A camera records the view as it travels away from the earth, each time multiplying the distance away by a factor of 10. The first frame is from 1 metre away, then 10 metres, 100 metres, 1 kilometre, 10 kilometres, 100 kilometres and so on. By only the 24th frame, the camera is billions of light years distant and the entire Milky Way galaxy is an invisible speck. If we use a simple arithmetic scale to measure this distance, the numbers get unmanageably large very rapidly. However, if we use a multiplicative scale, where each unit is a *multiplier of* the previous one, rather than an *addition to* it, the numbers can easily express the scale involved. We use *powers* to make the notation even more succinct. Thus, for example, instead of saying the camera is 1,000,000 metres away, we express this number as the number of times 10 has to be successively multiplied to reach it: 10^6 (pronounced '10 to the power 6') is just $10 * 10 * 10 * 10 * 10 * 10$. This is the same as the number of 0 digits after the '1', or the number of steps the camera has taken.

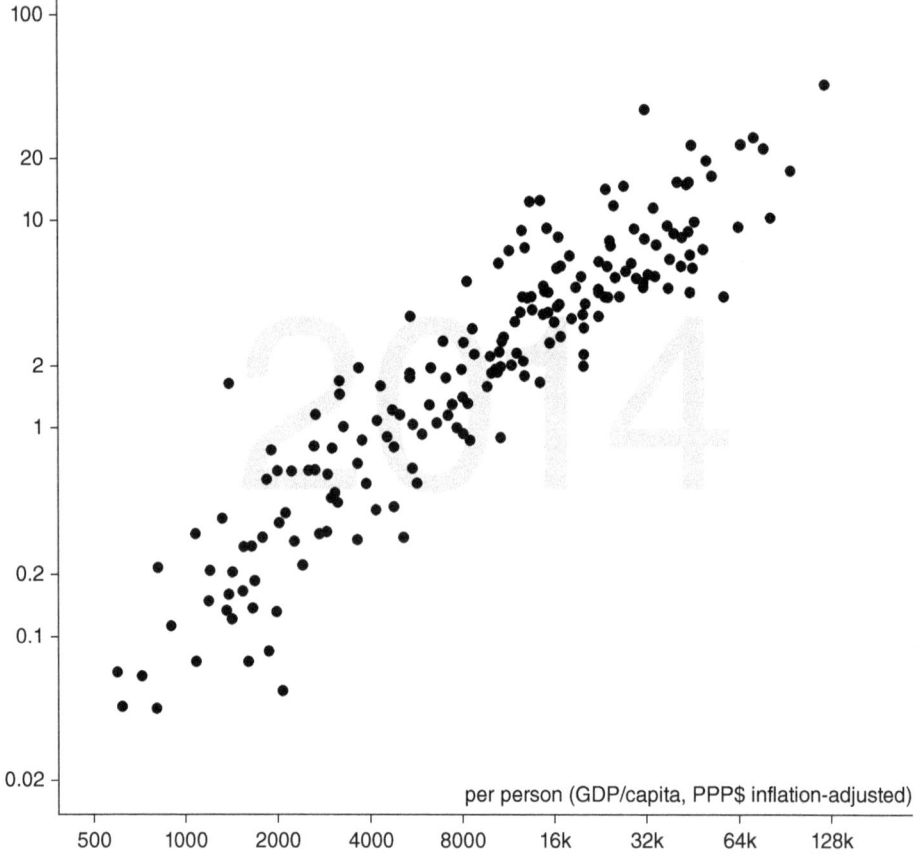

Figure 5.13 Mean income and carbon dioxide emissions (log scales)

Note. www.gapminder.org

We call the number that is multiplied by itself the *base*, and the number of times it is multiplied by itself as the *power*, sometimes also called the *exponent* or *index*. Double asterisks or a caret (^) are often used to indicate powers, or the power is shown as a superscript to the base. Thus, if the *base* is 2 and *power* is 5, we have

$$2 * 2 * 2 * 2 * 2 = 2^5 = 2\textasciicircum5 = 2**5 = 32$$

If the *base* is 10 and *power* is 4,

$$10 * 10 * 10 * 10 = 10^4 = 10\textasciicircum4 = 10**4 = 10{,}000$$

Arithmetic rules of powers

If two numbers expressed as powers to the same base are *multiplied*, this can be done by *adding* the respective powers.

Example: $10^3 * 10^2 = 10^5$ [because $(10 * 10 * 10) * (10 * 10) = 10 * 10 * 10 * 10 * 10$]

Example: $2^4 * 2^2 = 2^6$[$(2 * 2 * 2 * 2) * (2 * 2)= 16 * 4 = 64$]

Thus, we have the general rule: $a^m * a^n = a^{(m+n)}$.

Similarly, if two numbers expressed as powers to the same base are divided, this can be done by *subtracting* the respective powers.

Example: $10^3/10^2 = 10^1$ [because $(10 * 10 * 10)/(10 * 10) = 10$]

Example: $2^3/2^2 = 2^1$[$(2 * 2 * 2)/(2 * 2)= 8/4 = 2$]

Thus, we have the general rule $a^m/a^n = a^{(m-n)}$.

From this it must also follow that

- Any base number to the power 0 is equal to 1: $10^0 = 1$.
- Any base number to the power 1 is just the base number itself: $10^1 = 10$.

Powers can be negative

Negative powers express the *reciprocal* (that number by which the original number must be multiplied to obtain 1). The rules of addition/subtraction still apply (indeed we arrive at the meaning of negative powers from the rules of addition and subtraction outlined above).

Example: $2^{-2} = 1/2^2 = ¼$

Example: $2^3 * 2^{-2} = 2^1$[$(2 * 2 * 2) * (1/4)= 8 * 1/4 = 2$]

Thus, we have the general rules: $a^{-m} = 1/a^m$ and $1/a^{-m} = a^m$ and $1/a = a^{-1}$.

Powers need not be whole numbers

From the symmetry of multiplication and division, and from the addition rule it must also follow that

$$a^{n/m} = (a^{1/m})^n = (a^n)^{1/m}$$

Therefore,

$$9^{1/2} * 9^{1/2} = 9^1 = 9$$

$9^{1/2}$ = the square root of 9 = 3 (the number multiplied by itself that equals 9).

Therefore, the *n*th *root* of a number *x* is a number *a*, which, when raised to the power of *n*, equals *x*.

Example: $16^{1/2} = 4$ the second (square) root of 16 [because $4^2 = 4 * 4 = 16$]

Example: $16^{1/4} = 2$ the fourth root of 16 [because $2^4 = 2 * 2 * 2 * 2 = 16$]

Logarithms

The *logarithm* of a number to a *base* is the *power* to which that base must be raised to express the original number. For example, the log of 1000 to base 10 is 3, since $10^3 = 10 * 10 * 10 = 1000$. The log of 10,000,000 to base 10 is 7, because $10^7 = 10,000,000$. The log to base 10 of 10,000 is 4, because $10^4 = 10,000$. The base is usually written as a subscript so that we would write

$$\log_{10}(1000) = 3$$

To re-express a log as the original number we take the *exponent* of the log. Since the log is the power to which the base has been raised to express the original number, the exponent of a log is given by raising the base to that power.

$$\exp_{10}(3) = 10^3 = 1000$$

Logs are often useful because they represent relations of multiplication by addition. Thus, adding the logs of two numbers gives the same result as multiplying the original numbers. For example,

$$\log_{10}(1000) + \log_{10}(10,000) = 3 + 4 = 7$$

$$\exp_{10}(7) = 10^7 = 10,000,000 = 1000 \times 10,000$$

This is why, as you may have noticed, the log to base 10 of the numbers we have used equals the number of zeros after the 1, since our counting system also uses a base of 10. We often use logarithmic transformations when describing processes that are multiplicative rather than additive in nature, or where we have a distribution that is heavily skewed.

In practice, rather than use logs to base 10, we use *natural* logarithms which are to base *e* (= approximately 2.718). The term *e* denotes a number with various very desirable mathematical properties that is useful for describing processes of growth or decay that are exponential in form. Imagine a process in which the value of some-thing doubles in each time period as shown in Table 5A.1.

Table 5A.1 Growth represented by 2^{period}

Period	0	1	2	3	4	5	6	7	8
Value	1	2	4	8	16	32	64	128	256

We could express the value as a function of the time period by raising the number 2 to a power equal to the number of the period. For example, $16 = 2^4$. However, a little reflection will show that the 2 we are using as a base here comprises *two* numbers: the original value (1) and the amount of growth (100% = 1). We could turn this into a general formula for the rise in any value subject to a constant process of growth over discrete periods of time:

$$\text{Growth} = (1 + \text{rate of growth})^{period}$$

However, growth may not happen in this way. The key word in our definition was *discrete*, so that growth is imagined to occur in a finite series of distinct steps. An analogy would be a bank account in which interest was applied on the amount in the account at the end of each year. If you had such an account, deposited $100 and left it there, and the interest rate was 100%, you would have $200 at the start of year 2, $400 at the start of year 3 and so on. However, if the interest was applied more fre-quently (every month, or every day, or every second or fraction of a second) growth would be larger, since interest would start to accrue earlier on both the principal and the interest gained so far. However, there would be a limit to this process, which equates to the concept of growth being *continuous*. Using calculus, we can calculate this limit, and it turns out to equal the value of *e*.

The logit

The logit is the natural log of odds (i.e. log to base *e* of odds). As we have seen, odds can take a value between 0 and infinity, with the value 1 representing a probability of 0.5, values less than 1 representing lower probabilities and values greater than 1 representing higher probabilities. The natural logarithm of odds will take a negative value when the odds are less than 1 (i.e. probabilities less than 0.5) and a positive value when the odds are greater than 1 (i.e. probabilities more than 0.5). The logit of odds of 1 (equally likely outcomes) will be 0.

Activity: Key Terms

As usual, this chapter has covered a lot of ground. You can check your understanding by trying to define the following terms. If you find it difficult, you can easily use the index of terms at the end of the book to locate where each is discussed.

beta coefficients

between-group variance

coefficient of determination or R^2

conditional variance

correlation coefficient

covariance

dummy variable

intercept

logarithm

mean square between groups

mean square within groups

OLS regression line

regression sum of squares

residual sum of squares

slope

sum of squares

total sum of squares

transformation

unconditional or marginal variance

variance

within-group variance

Chapter Summary

- The F-statistic in ANOVA is the ratio of unconditional variance between groups defined by a categorical variable to the conditional variance within them.
- Multiple regression examines the variance in a response or dependent variable that can be accounted for by any categorical or continuous independent variable, while

controlling for the value of other independent variables with which the dependent is correlated, provided that a linear model of association is appropriate.

- Categorical independent variables with k categories can be represented by a series of $k - 1$ dummy variables.
- The ordinary least square regression line, described by the regression equation, minimises the sum of the squared residuals from it to the data coordinates of the observed values of the independent and dependent variables.
- Standard errors, p-values and confidence intervals can be calculated for individual regression coefficients and for the regression model.
- Powers and logs can often transform variables so that their associations become linear in form.

Further Reading

By far the best introduction to regression is Paul D. Allison (1999). *Multiple regression: A primer.* Pine Forge Press. It is a short book that does exactly what it says. Donald J. Treiman (2009). *Quantitative data analysis.* Wiley, contains a thorough treatment of regression and correlation.

6

POWER, EFFECT SIZE AND INVERSE PROBABILITY

Chapter Overview

Introduction

Chapters 3 to 5 have all described the logic of NHST. NHST is a powerful logic of evidence interpretation because it assumes no prior knowledge of what is under investigation. Such logic can be essential for a scientific approach. Many spectacularly counter-intuitive insights into the natural and social world have depended upon this ability to follow only where the particular piece of data under investigation appears to lead.

However, starting from scratch like this limits the range of inferences we can make. When we did our NHST trials of Coke and Pepsi in Chapter 3, we made *no* assumptions about *how skilled* cola tasters might be. Maybe those able to discriminate had no difficulty doing so and would usually get all 10 trials correct. Maybe even the best tasters had a success rate of only 60% or 70%. Neither did we make any assumptions about how *widespread* the ability to discriminate cola might be. Maybe nearly everyone had it, maybe no one. This did not matter to us, since all we did was set up a test that those with no skill were unlikely to pass. We compared the data we observed to the data we would have expected to observe under conditions of our 'no skill' null hypothesis, and if it was sufficiently different, we decided we had provisional evidence of skill. Our significance level, alpha (α), was our Type I error rate, which was equal to the long run proportion of those guessing randomly lucking out and scoring 8 or more out of 10. Our sampling distribution told us the probability of observing our data, *conditional upon the null hypothesis being true*. We could also think of our Type I error rate as our *false positive rate*, that is, the proportion of all the real negatives (= cases for which the null hypothesis was actually true) falsely reported as positive.

$P(\text{data} \mid H_0 = \text{true})$

Often, however, we might want to know something rather different: the probability of a hypothesis being true, *given the data we observe* and, associated with that, the *proportion of reported positives that are truly positive.*

$P(H_0 = \text{false} \mid \text{data})$

This is a vital distinction, and one worth taking some time to digest. Because we knew nothing of the strength of cola-tasting skill, or how widespread it was, we could have no idea what proportion of those who failed our tests were indeed without skill, or what proportion *of those who passed* our 8 from 10 threshold did so because they were lucky rather than having the skill. We thus did not know anything about our *false negative* rate, or the *power* of our test. Perhaps *nobody* can actually tell the difference between colas, and those who pass our test are all just lucky. Conversely, maybe many who failed our test actually *did* have some powers of discrimination – just not enough to be correct as many as eight times out of 10.

We saw in Chapter 3 that failing to reject a null hypothesis did not necessarily mean accepting it, far less proving it, even provisionally, and that just as we might commit a Type I error of rejecting a null hypothesis that was in fact true (i.e. mistaking noise for a signal), we could also make the complementary Type II error of failing to reject a null hypothesis that was in fact false (i.e. not detecting a signal in the noise). If we have good reason to think that our probability of making a Type II error is low, we might want to treat an un-rejected null as important evidence that the null was in fact true. This is to say, if our test has enough *power*, we can use it to provisionally confirm the null and decide that the absence of evidence *for* an effect does indeed constitute evidence that the effect *does not exist* in the population, not just that our test had been unable to detect it.

This might seem like splitting hairs, but keeping the limitations of NHST in mind is often essential. Imagine a drug trial, or a pilot for some social policy measure. For all kinds of reasons, the number of participants may have to be kept low. This inevitably reduces the power of any test, since the role of random variation must be relatively greater. In such circumstances, it could be dangerous to conclude that something was of no interest because it failed to reach significance. For example, we might have some evidence of serious side effects, or other unanticipated consequences. Such evidence might not be enough to *reject* a null hypothesis of 'no side effect', but it could be very dangerous to conclude that we had evidence *confirming* the null. This is not an imaginary scenario but what has actually happened in many different contexts (Ziliak & McCloskey, 2008).

We also need to remember in NHST that our significance level is also our long run Type I error rate. There has been a long interest in psychology in investigating the existence of paranormal phenomena, such as telepathy, premonitions, the impact of prayer and so on. Every so often, researchers come up with a result that is 'significant'. We would be foolish if we allowed such results to convince us. Given a traditional alpha level of 5%, we should expect about one in 20 of such experiments to be successful. My favourite in this genre are studies of the retrospective impact of effect of intercessory prayer demonstrating that prayers offered *after* the patients had recovered had an impact on the speed and quality of their recovery. Both these errors arise from the over-interpretation of one-off results as significant or insignificant. A significant result is not 'proof', but rather a signal that merits further investigation. Conversely, failure to reject a null hypothesis is not 'proof' that nothing is there, unless we are confident that our test has been powerful enough to detect effect sizes that are not trivial.

Thomas Bayes: prior and posterior probabilities

In order to examine the *power* of tests, we need some information about what we are trying to measure. This inevitably violates the assumption of 'no prior knowledge', but sometimes we do already have some evidence about what we are investigating, and this makes it possible to approach any new evidence we have in a different way. We can summarise our existing knowledge in terms of one or more *prior* probabilities or base rates for a phenomenon and then infer how much any new knowledge we encounter ought to lead us to revise this probability, creating a *posterior* probability that takes account of our new knowledge. We can use similar calculations to decide what the *false negative* probability, *Type II error rate* or *power* may have been in any inference we make.

This use of probability is something that you may already do informally, albeit in an unsystematic way. As gregarious social animals, we tend to make character judgements about new acquaintances very rapidly. Someone may strike us as more or less pleasant or pushy, playful or boorish, critical or credulous and so on. These first impressions (= prior probabilities) may later be reinforced, revised or even reversed (= posterior probabilities) as we become more familiar with that person. As we accumulate more evidence, our opinions can change. However, working in this way also illustrates a potential risk of working with prior probabilities. If they are badly mistaken to begin with, it may take some time for us to realise our error. Worse, as we saw in Chapter 1, we have a cognitive bias towards interpreting new evidence in a way that is as consistent as possible with our existing beliefs and prone to discount conflicting evidence. Rather than correcting our original bias, we may simply make it yet more extreme.

Taking a prior probability and using new evidence to produce a posterior probability is one application of Bayes's rule, named for Thomas Bayes, who set out his views on inverse probability in an essay read to the Royal Society after his death by his friend Richard Price, who edited and published it. Much of Bayesian reasoning leads to voluminous calculations that were not practicable until the invention of cheap computing in the 1970s, but the period since then has seen an explosion of interest in this approach. The 'frequentist' logic we examined in Chapters 2 to 5 defined probability as the limit of the frequencies in a probability distribution created by a long run of repeated identical trials. Bayesian approaches use a quite distinct definition of probability as a *degree of reasonable belief* in a piece of evidence, prediction or description (which nevertheless shares a lot of the mathematics with frequentism). Bayesian approaches can be very useful in situations where the concept of a long run is irrelevant or meaningless – for example, in estimating the probability of very infrequent or unique events. However, critics of Bayes argue that the concept of a prior probability can sometimes be inconsistent with the scientific aim of emancipating knowledge from existing belief, and even suggest that the reason Bayes did not publish his great work while he was alive was because of his own reservations about it.

To get an insight into the logic of Bayesian reasoning and the calculations it makes possible, let's return for a final run of Coke and Pepsi trials. First, let us imagine we already had some evidence that there was an 'all or nothing' contrast between those with the skill and those without it, and that expert cola drinkers scored an average 9 out of 10 glasses right. This is a substantial effect: a score of 9 compared to 5 on average for a pure guesser. We could now construct a sampling distribution for k correct identifications under the hypothesis that the long run probability of success in the trials was 0.9 for a skilled cola discriminator, using the binomial formula as we did before. Figure 6.1 shows the results.

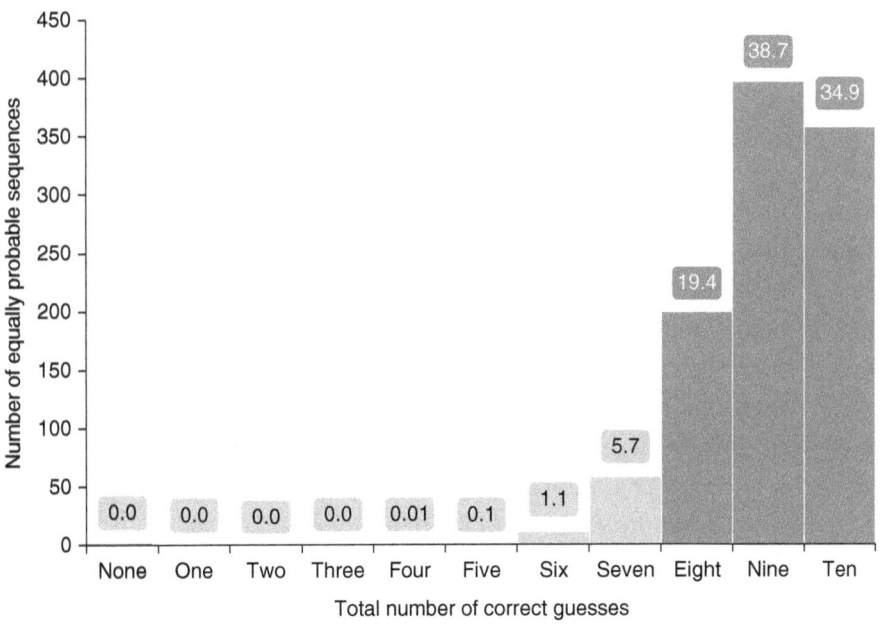

Figure 6.1   Sampling distribution for $p = 0.9$

If we kept to our previous criterion of eight or more out of 10 correct identifications as evidence of skill, only about 6.9% (5.7 + 1.1 + 0.1) of those who really did have the skill would fail our test. This is our *false negative rate*, also known as the *Type II error rate or beta (β)*. It is calculated by the following formula:

$$\text{False negative rate} = \frac{\text{False negtives}}{\text{True positives} + \text{False negatives}}$$

The *power* of a test is given by subtracting this amount from one, so that here,

Power $= (1 - \beta) = (1 - 0.069) = 0.93$

That looks like a fairly powerful test. The *power* of a test is defined as the probability of rejecting the null *when it is in fact false*. Would we therefore be safe to go beyond 'failing to reject' the null of no skill for that 6.9% of subjects and positively conclude that they did have no skill, confirming the null for them? Not quite yet!

Keep in mind that the sampling distribution shown in Figure 6.1 is only for those who *do* have the skill. Type II error rate (β) calculations give us the probability of rejecting the null, conditional upon the null hypothesis being false. We were able to calculate it because we made an assumption about the *effect size* we were trying to capture (= 9/10 correct identifications). It tells us the proportion of all *real positives* falsely reported as negatives. But what we may really want to know is the proportion of *all* our *reported* negatives that were true ones. To produce a sampling distribution for the whole population, rather than one for those with skill (Figure 6.1) and one for those with no skill (Figure 3.3), we would need to know *what proportion of the population had the skill* in the first place, or the prevalence of the signal we are trying to detect, as well as its size. If a large proportion of the population did not have the skill, then we'd be fairly safe in taking a failure to identify at least eight drinks correctly as evidence that the null was true. But what if, say, only 5% of the population were unable to discriminate between colas? Fifty people in every 1000 would not have the skill, and because our test is accurate, 47 of them would be correctly classified as true negatives (from the sampling distribution in Figure 3.3). The remaining 950 people in every 1000 would have the skill, and while 884 of them would be correctly classified, 66 would be false negatives (from the sampling distribution in Figure 6.1). Thus, although we had a test that was powerful enough to identify cola discrimination skill accurately when it was present (93% of the time) and to rule it out when it was absent (94.5% of the time), because the prevalence of the skill in the population was very high, most of the people *failing* our test would be people *with* the skill!

This apparent paradox arises from confusing two fundamentally different pairs of probabilities. Type I and Type II error rates, α and β, significance and power, all express false positives as a proportion of *all true negatives* and false negatives as a proportion of *all true positives*. Given a sampling distribution for true negatives (our null hypothesis), we can work out the alpha (α) and the Type I error rates. Given a sampling distribution for true positives (our hypothesis of $p(\text{correct}) = 0.9$) we can work out the beta (β) and the Type II error rates. However, we often wish to know what proportion of all *reported positives* are true positives and what proportion of all *reported negatives* are *true negatives*. That is to say, we wish to know *the probability of our hypothesis given the evidence*, rather than *the probability of the evidence given our hypothesis*. As we have seen, we can only make this latter calculation if we have a *prior probability* of the hypothesis being true, established by some evidence about the distribution of the phenomena we are trying to measure in the population such as its size and prevalence.

This might look like inference eating its own tail. Isn't it because we *cannot* measure the population that we are calculating inferences in the first place? The key issue is the range of evidence available to us, and given that, the kind of inferences we can safely make. In a situation where we have no good prior empirical knowledge, NHST is as far as we can go. As we have seen, it tells us how likely we are to witness some result if our null hypothesis model of the state of the population is true. If that probability is low enough, and out test was powerful enough, we have provisional evidence that *something* is going on, but we will almost certainly want to follow that up with further analysis to determine what that something might be. Our 'result' is likely to be the beginning of the research process rather than its conclusion. We know $P(\text{data} \mid H_0)$ but we are far from knowing $P(H_0 \mid \text{data})$, so that we cannot know if our result was likely to be one of the *reported* positives that was also true, or what proportion these might take in the long run. However, where we have other knowledge, and have reason to believe it is fairly secure, we can use that to extend the range of inferences we can draw.

In my experience, our brains struggle to keep these distinct sets of probabilities clear. Gigerenzer and Hoffrage (1995) argue, correctly in my view, that everything is made much clearer by working with raw frequencies and a contingency table of the form shown in Table 6.1. I urge you to spend some time examining this table, together with Tables 6.2 and 6.3, as they will make any textual description of probabilities and inference much clearer.

Table 6.1 Trials and results: raw cases

		Real Parameter		
		H_0 TRUE	H_0 FALSE	
Trial result	**Reject H_0**	*False Positives* *(reject true null hypothesis)* *Type I error*	**True Positives** *(reject false null hypothesis)* **CORRECT**	**All reported positives** **'success'** **'result'**
	Do not Reject H_0	**True Negatives** *(accept true null hypothesis)* CORRECT	*False Negatives* *(accept false null hypothesis)* *Type II error*	**All reported negatives** **'failure'** **'no result'**
		All real negatives	**All real positives**	**All tests**

Table 6.2 Trials and results: column proportions

		Real Parameter		
		H_0 **TRUE**	H_0 **FALSE**	
Trial result	**Reject** H_0	Type I error rate α P(result | H_0 TRUE)	SENSITIVITY P(result | H_0 FALSE) $1 - \beta$ Power	All reported positives 'success' 'result'
	Do not Reject H_0	P(no result | H_0 TRUE) SPECIFICITY $1 - \alpha$	P(no result | H_0 FALSE) Type II error rate β	All reported negatives 'failure' 'no result'
		All real negatives	All real positives	All tests

Table 6.3 Trials and results: row proportions

		Real Parameter		
		H_0 **TRUE**	H_0 **FALSE**	
Trial result	**Reject** H_0	P(H_0 TRUE | result)	P(H_0 FALSE | result)	All reported positives 'success' 'result'
	Do not Reject H_0	P(H_0 TRUE | no result)	P(H_0 FALSE | no result)	All reported negatives 'failure' 'no result'
		All real negatives	All real positives	All tests

The building block of probability, as we saw in Chapter 2, is the Bernoulli trial, which can result in success or failure. To keep our language consistent, we can think of 'success' in a trial as a positive result, such as the skill of correctly identifying the type of cola in a glass. We can imagine either individual trials or series of them, such as our 10-glass trial in which a score of 8 or more was taken as 'success'. The results of our trials, which we do observe, are attempts to describe population parameters that are themselves unobserved by means of some hypothesis, usually a null hypothesis whose *rejection* would give us some information about the population. Any individual trial in which a null hypothesis is tested faces four possibilities, corresponding to each of the cells in the body of Table 6.1:

> The null hypothesis is a true description of the population, but our trial gives a positive result and causes us to reject the null hypothesis, so that we commit a Type I error.
>
> (top left cell)
>
> The null hypothesis is a true description of the population, our trial gives a negative result, we do not reject the null hypothesis, so that we are *correct*.
>
> (bottom left cell)

The null hypothesis is a false description of the population, our trial gives a positive result and causes us to reject the null hypothesis, so that we are *correct*.

(top right cell)

The null hypothesis is a false description of the population, our trial gives a negative result, we do not reject the null hypothesis, so that we commit a Type II error.

(bottom right cell)

Table 6.2 re-presents Table 6.1 focusing on analysis of the conditional probabilities that are shown in the *columns* of the table, or the probabilities of observing our data, *conditional upon* whether our hypothesis is true or false. NHST procedures require only the information in the left-hand column of our table. We need no prior information on the population parameter we are trying to observe, either how substantial it is or what its prevalence might be. We could describe the contents of the column as the sampling distribution for our null hypothesis. We could set a Type I error rate or α or significance level that we are prepared to accept and, having done so, reject the null hypothesis when the data we observe is sufficiently improbable when the null is true. If we wish to draw conclusions about the *power* of our test, we can calculate how big the effect size would need to be in the population before our tests pick it up using information in the right-hand column of Table 6.2. To continue our example of cola discrimination, we would need to estimate how good cola discriminators tended to be to know what proportion of people with the skill were nevertheless likely to fail our test. In a medical context, the terms *specificity* and *sensitivity* are often used.

Specificity is the probability of accepting the null when it is indeed true, or obtaining a negative rather than a false positive result. The Type I error rate (which is also the false positive rate) and specificity of any test must sum to one, since they express the division of all true negatives into those which are correctly and incorrectly classified by our test:

$$\text{Specificity} = \frac{\text{True negatives}}{\text{True negatives} + \text{False positives}} = 1 - \alpha$$

Sensitivity is the probability of rejecting the null when it is indeed false or obtaining a false positive rather than a false negative result. The Type II error (which is also the false negative rate) and sensitivity must sum to one since they express the division of all true positives into those which are correctly and incorrectly classified by our test:

$$\text{Sensitivity} = \frac{\text{True positives}}{\text{True positives} + \text{False negatives}} = 1 - \beta$$

Table 6.3 shows the conditional probabilities in the *rows* of our contingency table, or the conditional probabilities of the correctness of our results, conditional upon the data we observe. The calculation of these probabilities requires us to have some estimate of the truth of the original hypothesis in the population, or the marginal probability that is $P(H_0)$. If we have an estimate of that, we can use that information to estimate the probability that any result we obtain is actually a correct one. To do so, we need the row probabilities given by:

$$\frac{\text{True negatives}}{\text{True negatives} + \text{False negatives}}$$

That is, the probability that the result is correct, given that the result is negative; and

$$\frac{\text{True positives}}{\text{True positives} + \text{False positives}}$$

That is, the probability that the result is correct, given that the result is positive.

The row and column conditional probabilities in a contingency table are related in a formula named for Bayes:

$$P(H_0|\text{ data}) = \frac{P(\text{data }|H_0)*P(H_0)}{P(\text{data})}$$

You will often see this formula written in the more general form of

$$P(A\,|\,B) = \frac{P(B\,|\,A)*P(A)}{P(B)}$$

If you examine the formula, you will see that it expresses the conditional row probability as a function of the conditional column probability and the two marginal probabilities in the table.

All this is much easier to see by working through a couple of examples, so let us return for the last time to Coke and Pepsi. First, we can populate our table with the raw frequencies in the example we last considered. We imagined a scenario in which 95% of the population could discriminate between colas with an average score of nine out of 10 correct identifications. The remaining 5% of the population had no skill. What would happen when we administered our test to 1000 subjects? We can fill in the bottom marginal row probability with our knowledge that for 95% of the population the H_0 of 'no skill' is false. From the sampling distribution we calculated earlier shown in Figure 3.3, for the 50 subjects with no skill, setting our alpha level at 5.5% by requiring the correct identification of eight colas means that about three of those with no skill will get lucky and pass our test even though they have no skill. We will correctly classify the remaining 47. From the sampling distribution shown in Figure 6.1 for the subjects who do have the skill, we know that about 884 of them will pass our test, while 66 will fail it. Overall, 887 will be defined as skilled cola tasters by our test while 113 will fail it, as shown in Table 6.4.

Table 6.4 Cola discrimination test (95% of population with skill)

		Real Parameter		
		H_0 **TRUE (No Skill)**	H_0 **FALSE (Has Skill)**	
Trial result	**Reject** H_0 8+ glasses correct	3	884	887
	Do not Reject H_0 ≤7 glasses correct	47	66	113
		50	950	100

Expressing these raw numbers as marginal and conditional probabilities, we can check Bayes's rule and also see how it expresses the row conditional probability of 47/113 in terms of the column conditional probability (47/50) and the two marginal probabilities (50/1000) and (113/1000):

$$\frac{47}{113} = \frac{\frac{47}{50} * \frac{50}{1000}}{\frac{113}{1000}} = \frac{\frac{47}{50} * \frac{\overline{50}}{\overline{1000}}}{\frac{113}{\overline{1000}}}$$

In our example, we were able to estimate the probability that a result was correct because we had a prior probability for getting such a result in terms of our information about the prevalence of cola discrimination skills in the population, and the sensitivity and specificity of our test. Of course, this does not tell us *which* of the 113 subjects who failed our test were the 66 false negatives and which were the true negatives. Our conditional probability describes the negative results as a whole, not individual members. Note, however, how much administering our test has improved our knowledge. Prior to its administration we'd need to assume that any individual picked at random had a 95% probability of being a cola discriminator. After taking the test, were they to pass it, we could be 884/887 = 99.7% confident that they had this skill. After failing the test, as we have seen the probability that they were a cola discriminator would fall to 66/113 = 58.4%. Were we to hold fast to the motto of 'nothing on another's word' and discount our prior probability, that is, were we to assume no knowledge of the extent of cola discrimination skill in the population, our best estimate would be only the result of our test, which would have declared 88.7% of the population to have the ability.

Now imagine a situation where half the population can discriminate between the colas, again with an average score of 9 out of 10, as in the sampling distribution in Figure 6.1. How well would we do in classifying them now? As before, we can populate our contingency table. If we examine 1000 people, we expect around 500 will have the skill and 500 will not.

Of the 500 *with* the skill, 0.069 * 500 = 35 will be incorrectly classified as *not* having the skill or *false negatives* (corresponding to the light grey area in Figure 6.1) while 500 − 35 = 465 will be correctly identified or *true positives*. As we saw back in Chapter 3, of the 500 without the skill, 0.054 * 500 = 27 will be incorrectly classified as having the skill or *false positives* (corresponding to the dark grey area in Figure 3.3) while 500 − 27 = 473 will be correctly identified or *true negatives*. Overall our test will have correctly identified 938 people, and misclassified 62, as shown in Table 6.5.

Table 6.5 Cola discrimination test (50% of population with skill)

		Real Parameter		
		H_0 TRUE (No Skill)	H_0 FALSE (Has Skill)	
Trial result	**Reject** H_0 8+ glasses correct	27	**465**	**492**
	Do not Reject H_0 ≤7 glasses correct	473	35	508
		500	**500**	**100**

Given our assumptions about the nature of the cola skill and its prevalence in the population, our test does well, classifying 94% of people correctly. This is because we have a strong signal (discriminators score 9/10 on average, compared to guessers), and it is widespread in the population. What happens if the signal is weak: let's have discriminators score 7 on average. The sampling distribution for discriminators will now be defined by the binomial distribution for $p = 0.7$, $n = 10$, shown in Table 6.6.

Table 6.6 Sampling distribution for k, n = 10, p = 0.7

N correct glasses	0	1	2	3	4	5	6	7	8	9	10
P		0.00	0.00	0.00	0.01	0.04	0.10	0.20	0.27	0.23	0.12	0.03

Thus, only 38.2% of cola discriminators will now pass our test, and our crosstab will look like Table 6.7.

Table 6.7 Cola discrimination test (50% of population with skill; weak signal)

		Real Parameter		
		H_0 TRUE (No Skill)	H_0 FALSE (Has Skill)	
Trial result	**Reject** H_0 8+ glasses correct	27	191	218
	Do not Reject H_0 ≤7 glasses correct	473	309	782
		500	**500**	**1000**

Because of the weak signal, our test now does poorly. Two out of five of those failing our test will actually have some skill, and overall our test will give the correct result only about two times out of three.

Taxi for Kahneman and Tversky

The fact that the low prevalence of an effect in a population causes a surfeit of false positives unless the power of a test is high is examined in a famous puzzle first proposed by Daniel Kahneman and Amos Tversky as a dramatic introduction to Bayes's rule (Hacking, 2001). In a town there are two taxi companies, *Green Cabs Ltd.* and *Blue Taxi Inc.* Blue Taxi uses cars painted blue; Green Cabs uses green cars. Green Cabs dominates the market, with 85% of all taxis. On a misty night, a taxi collides with another car and drives off. A witness to the collision says they saw a *blue* cab. The witness is tested under conditions like those on the night of the accident, and 80% of the time they get the colour of the car they see correct. That is, regardless of whether they are shown a blue or a green cab in misty evening light, they get the colour right 80% of the time. What can we conclude, on the basis of our witness evidence? The temptation is to conclude that because the witness is right 80% of the time, the probability of the taxi being blue is 80%. Kahneman and Tversky argue, counter-intuitively, that the balance of probability is that the taxi was green! Taking *only* the witness statement ignores the evidence we also have about the *prevalence* of green taxis in the town. Given the information that the witness gets the colour right 80% of the time, we can populate our contingency table as follows for testing the witness with 1000 random taxi sightings (Table 6.8).

Table 6.8 Witness performance at 80% success rate

		Real Parameter		
		Green	Blue	
Witness sees	Blue	170	120	290
	Green	680	30	710
		850	150	1000

Why? Because of what Kahneman and Tversky call the *base rate* or we have described as the 'prior probability' of colour of the taxi *before* the witness saw it. The key factor to consider is that there were so many more *green* than blue cabs on the streets. Because of this, any sighting of a 'blue' cab *is more likely to have been a misrecognised green cab (170/290 =59%) than a correctly identified blue cab (120/290 = 41%).*

We can also work through this in terms of marginal and conditional probabilities, using Bayes's rule.

Here is Bayes's rule again:

$$P(H|\text{Evidence}) = \frac{P(H)*P(\text{Evidence}|H)}{P(H)*P(\text{Evidence}|H)+P(\text{Not }H)*P(\text{Not}H\,|\text{Evidence})}$$

We want to know the probability of the **hypothesis** '*the cab is blue*' conditional upon, or given the **evidence** '*the witness sees a blue cab*'. That is **P(H | evidence).**

What is the 'prior' probability that the hypothesis is true? It is the probability of any cab being blue, regardless of what the witness thought = 0.15 (since 15 out of 100 cabs on the road are blue). This is **P(H).**

What is the probability of the evidence conditional upon a true hypothesis?

That must be the probability of the witness seeing a blue cab when a cab is indeed blue = 0.8. This gives us **P(Evidence|H).**

Therefore, the top line of the fraction in our equation will be 0.15 * 0.8.

Note that the first part of the bottom line of the fraction P(H) * P(Evidence|H) is just the same as the top line. The second part is the probability of the hypothesis not being true multiplied by the probability of the hypothesis not being true, given the evidence.

The prior probability of the hypothesis not being true is simply the probability of the cab being green = 0.85 (since 85 out of a 100 cabs are green). We could also get this by subtracting P(H) from 1. Remember that the hypothesis (cab is blue) must either be true or untrue (all cabs are blue or green), so if 0.15 of cabs are blue 1 − 0.15 = 0.85 of cabs must be green. This gives us **P(Not H).**

What is the probability of a cab being green (not H), given the evidence 'the witness sees a blue cab?' It is the probability of the witness misrecognising a green cab for a blue cab, which we know is 0.2. So **P(Not H|Evidence)** = 0.2.

Thus, we can now put numbers into our equation:

$$P(H\,|\,\text{Evidence}) = \frac{0.15 * 0.8}{(0.15 * 0.8) + (0.85 * 0.2)} = \frac{0.12}{0.12 + 0.17} = \frac{0.12}{0.29} = 0.414$$

If you look back at the contingency table, you will see that all the equation does is compute the conditional probability of a cab being blue when the witness thought it was blue from the prior distributions of the hypothesis determined by the colour of the cabs, and the probability of a correct identification.

We could sum up what we have just done by re-describing the equation as follows:

$$\frac{P(\text{Correct identification of blue cab})}{P(\text{Correct identification of blue cab}) + P(\text{Incorrect identification of blue cab})}$$

Be careful to note however that we are saying 'it is more probable' *not* that this must have been the case or would always be the case. The cab in the accident *definitely had a colour*. What we are uncertain about is what we can infer from our empirical evidence: the report of a blue cab by a witness (or test if you like) who on average can correctly identify the colour only 80% of the time (or on four out of five occasions). We can have no way of knowing whether *this particular occasion* was one of the four out of five correct identifications or the one out of five errors. But what we *do* know, and have calculated, is what the probability of a correct identification was, given *both* the probability of the witness making a correct identification *and* the 'base rate' probability of colour of the cab being identified.

This is analogous to the level of probability that we assign to confidence intervals for a sample statistic. The confidence interval either definitely contains or definitely does not contain the true population parameter we are trying to measure. The degree of probability, such as '95% confident' refers to our confidence in our procedure.

Our prior estimation of the probability of the cab being blue was 0.15 (the prevalence of blue cabs). The eyewitness testimony of the cab being blue increased our confidence in this eyewitness testimony, raising it to 0.41. However, the balance of evidence is still against the cab being blue, because the reliability of the eyewitness testimony was not high enough for us to conclude that a true sighting of a blue cab was more likely than a misrecognition of a more common green cab as a blue one.

Now imagine that our eyewitness had keener sight and was better at spotting the colour of cabs in the light conditions prevailing at the time of the accident. Let's see what happens if rather than being correct 80% of the time she was correct 90% of the time. Our equation would now be as follows:

$$\frac{0.15*0.9}{(0.15*0.9)+(0.85*0.1)} = \frac{0.135}{0.135+0.085} = \frac{0.135}{0.220} = 0.614$$

Our prior probability remains the same as in our first scenario, but its modification by the new evidence is greater, because that evidence is stronger, changing the balance of probability to correct identification of a blue cab, since there are fewer 'false positive' sightings of 'blue' cabs that are really green.

Conversely, imagine that the prevalence of green cabs on the road had been greater, so that instead of 85% of cabs they formed 95%. Our equation would then have been (keeping our eyewitness reliability at its original 80%)

$$\frac{0.05*0.8}{(0.05*0.8) + (0.95*0.2)} = \frac{0.04}{0.04+0.19} = \frac{0.04}{0.23} = 0.17$$

This time the lower initial probability (now only 5%) isn't raised much by the new evidence to make the posterior probability only about a 1 in 6 chance that the cab was blue.

A more substantial application: medical tests

This kind of logic is crucial when it comes to interpreting the results of medical tests. Just like any test, medical tests are not foolproof. Often we need to work out the implications of an individual test result, given the prevalence of a condition in the population, and given the long run probability of the testing procedure reaching a correct diagnosis. This can lead to some vital but counter-intuitive findings.

Medical tests, or other measurements or classifications, often have a high sensitivity and specificity, sometimes approaching 100%. Because of this, it is tempting to assume that because cases thus cluster along the diagonal of our contingency table in the True Positive and True Negative cells, the row percentages will resemble the column percentages so that true *results*, whether positive or negative, approach 100% of all results. This does not follow! The probability of any result being a true result will *also* depend on how observations are distributed between the columns of the table, which capture the prevalence in the population. If the prevalence is very low, which can often be the case when screening among the general population for a condition, then because so many more cases will be in the left-hand column of the table, a positive result will be more likely a false diagnosis of one of the many healthy patients, than a true diagnosis of one of the few genuinely ill ones! No matter how small the number of false positives is as a proportion of all the healthy patients tested (the column proportion), it is their share of all positive *results* that gives us the probability of the correctness of the test result (the row proportion). Finally, as we always do with inference, we need to keep in mind that this probability is a property of *all* the positive test results together, *not* a description of any single one. Any individual test result is either actually true or false, but that is not knowledge we have access to.

The key point is that, just as in the example with the taxi cabs, where the decisive factor was the prior probability, or base rate – the fact that there were so many more green than blue cabs – so here the decisive factor is the prevalence. If the vast majority of people undergoing a test do *not* suffer from the illness being tested, then even if the test is *very* sensitive and has a *low* false positive rate, false positives will still outnumber true positives, *because the latter will be so rare.* Even the best tests are not perfect and have a margin of error. Even a very small false positive rate will produce many more false positives than true positives if the prevalence of the condition in the population being tested is lower than the false positive rate in the test.

Covid-19

As the SARS-CoV-2 pandemic spread, different companies and others developed antibody tests to detect whether a patient had been infected with coronavirus disease 2019 (COVID-19). Public Health England's tests of the performance of one of these tests found it to have a specificity of 99.6% and sensitivity of 93.9%. At first, this looks like a promising performance. Out of every 1000 patients without COVID-19, only four would return a false positive result. Out of every 1000 patients who *did* have COVID-19, only 61 would return a false negative result. This level of sensitivity could create problems. Were the test to be used to screen healthcare workers who had a high risk of exposure to the disease, for example, it would give 6% of those who actually did have COVID-19 a false negative result. While the specificity looks extremely high, it would also cause problems if used for screening in populations where there was a very low prevalence of COVID-19. If about 0.4% of the population were infected, equivalent to around 280,000 people in a country the size of the UK, for example, a positive result in a mass screening programme would indicate only a 50/50 chance of actually having the disease. For every 1000 people tested, about four people would have the disease and test positive, while of the 996 free of the condition, about 4 would give a false positive result. Probabilities like these often make screening a poor substitute for quarantine, even with highly accurate tests. *Any* probability of a false negative risks releasing a contagious person into the general population, while low prevalence of the condition means that a significant proportion of positive results will be false.

Breast cancer screening

One recent estimate of the accuracy of mammograms is that their probability of detecting a cancer that is present is 0.83. Their probability of falsely indicating the presence of a cancer when it is not present is around 0.07. The prevalence of breast cancer is estimated to be around 1% for women aged between 50 and 70 in the UK. Let's use these figures to populate our Table 6.9.

Table 6.9 Mammogram result and real condition

		Real Patient Condition		
		Patient *does not have cancer* H_0 **TRUE**	Patient *does have cancer* H_0 **FALSE**	
Mammogram *result*	**Positive**	**693** **False positive**	83 *True positive*	776 Positive results
	Negative	9207 *True negative*	**17** **False negative**	9224 Negative results
		9900	100	10,000

Given the 1% prevalence of cancer, we could expect potentially 100 out of 10,000 women aged 50 to 70 to have breast cancer. If all 10,000 of these women came forward for screening, then 83 of the 100 women with breast cancer would be correctly diagnosed as having the condition, while 17 would receive false negative results.

Of the 9900 women free from breast cancer, 9207 would be correctly diagnosed as not having the condition, while 693 would get a false positive reading. Thus, although the mammography is 83% accurate, we could expect only $83/(693 + 83) = 11\%$ of the positive test results to be true positives. A positive mammogram is not necessarily bad news, although it may mean uncomfortable, distressing or invasive follow-up tests to rule out a cancer that was never there or treat one that would not have eventually proved malignant.

This performance is better than it might appear at first sight. What are the chances of *having* cancer if the mammogram is *negative*? It will be $17/9207 = 0.2\%$. We've used Bayesian reasoning to revise our expectation of the likelihood of cancer being present from a very low 1% (the prevalence rate) to a much higher 11% in the light of a positive result of the test and much lower 0.2% given a negative one. Were we to apply the logic of NHST only, and take no account of the prevalence of the condition in the population (perhaps because we had no good evidence of prevalence), we could go only by the results of our test, which would lead us to overestimate the prevalence of the condition. Finally, keep in mind that while we might be able to estimate the probability that a positive test result is a true positive, this tells us only about *all* positive test results, not whether any individual positive result is correct or not.

'Further Reading' section at the end of this chapter includes a recent report on the likely impact of cancer screening in the UK. An apparently *simple* question (does screening confer benefit?) informed by ample research data (RCTs [randomised controlled trials], meta-analysis, etc.) is nevertheless hard to answer, because although the benefits of screening can be estimated, albeit within a fairly wide range, the harm of misdiagnosis is by definition more difficult to measure. It is not usually possible to go back to a medical diagnosis and find conclusive proof about whether or not it was correct. Accordingly, we may often be unsure if a mastectomy, for example, represents a true cancer diagnosis and potential mortality averted, or a false diagnosis with attendant patient harm. Counterfactuals such as these are extremely difficult to estimate well. Inference enables us to understand and quantify uncertainty better. It cannot, on its own, transform uncertainty into certainty.

Chapter Summary

- NHST focuses on $P(\text{data} \mid H_0 = \text{true})$, but we often wish to know $P(H_0 = \text{true} \mid \text{data})$. This can be estimated if we have a *prior* probability for $H_0 = \text{true}$ that we revise using new evidence to calculate a *posterior* probability.
- If we have an estimate of the size of the effect or magnitude of the phenomenon we are investigating, we can estimate the *power* of our study to identify it.

- The false negative rate, Type II error rate or beta (β) = $\dfrac{\text{False negatives}}{\text{True positives} + \text{False negatives}}$
- The sensitivity or power of a study is $1 - \beta$.
- Raw frequencies displayed in a contingency table are the best method for considering Type I and Type II errors, false and true negatives and positives, specificity and sensitivity, inverse probabilities and Bayes's rule.
- When the prevalence of condition or the size of an effect is very low, even tests with high specificity and sensitivity may produce a high ratio of false positives to true positives.

Further Reading

An excellent description and discussion of informal Bayesian reasoning is given by Tetlock and Gardner in *The Art and Science of Prediction* (2015, Random House Books). Thorough introductions to Bayesian approaches to inference include Kruschke, J. (2014). *Doing Bayesian Data Analysis*, Second Edition. Academic Press and Gelman, A., Carlin, J. B., Stern, H. S., Dunson, D. B., Vehtari, A., Rubin, D. B. (2013). *Bayesian Data Analysis*, Third Edition. Chapman and Hall/CRC. Just how difficult it is to produce apparently simple measurements, such as the performance of mammography tests, is given by a recent independent review of screening in the UK (Independent Breast Screening Review, 2013) *The Benefits and Harms of Breast Cancer Screening: An Independent Review*.

7

WHAT DOES SOUND INFERENCE COMPRISE?

Chapter Overview

> How could it be, almost 100 years after the formulation of the Fisherian and Neyman-Pearson approaches to statistical inference, that a body as eminent as the ASA needs to step in and remind scientists how to define an index as ubiquitous in scientific investigations as the *p*-value and how to use it or other indices properly? And what will prevent us from dusting this same statement off 100 years hence to remind the community yet again of how to do things right? If we don't start focusing on that, the most likely outcome of this effort is that the statement will literally stand the test of time, in being as needed in the next century as it is today. (Goodman, 2016, p.1)

> [Banning NHST procedures] is like recommending decapitation as a headache remedy: it solves the problem but kills the patient. (Ziliak, 2016, p. 88)

If you have read and understood the chapters in this book, you should have a good grasp of the underlying logic of statistical inference. You should also be well equipped to understand and follow the vigorous debate that has developed over the last two decades over how statistical inference should best be used. Many critical articles have been published about the way in which *p*-values and significance thresholds are being used by scientists, peer reviewers and journal editors in different disciplines. Ziliak and McCloskey (2008) claimed that as many as 9 out of 10 articles in leading health, medicine, psychology and economics journals misused tests of significance. Often, it has been suggested that what has been called the 'reproducibility crisis' in science – the situation in which many studies, including some well-known ones, have failed to replicate when scientists have attempted to repeat the original research – is a function of the misuse of NHST procedures.

The ASA debate

In 2014, the ASA formed an expert group charged with drawing up a policy statement on *p*-values: the first time it had taken such a position on a matter of statistical practice rather than public policy. Developing the statement proved challenging. Six principles and a text enfolding them emerged from two years of intensive iterative debate by a large group of eminent statisticians and scientists. On publication in 2016, it was accompanied by commentaries from dozens of leading statisticians, including members of the working group, who each put their own gloss on the statement that had finally been agreed. In 2017, the ASA organised a symposium on inference, and in March 2019, it published an edition of its journal dedicated to articles from the symposium. Simultaneously, a letter to the prestigious journal *Nature*, signed by more than 800 statisticians and scientists, argued that it was 'time for statistical significance to go' (Amrhein et al., 2019, p. 307). The editors of *Nature*, probably the world's leading scientific journal, made clear where their sympathies lay by titling the accompanying editorial 'It's time to talk about ditching statistical significance'. That did not put an end to the matter, nor should it have done, since it is far from clear that the signatories represented a scientific consensus, or had a clear alternative to the NHST framework to recommend. Many others might share the criticism they advanced of the use of NHST but draw very different conclusions about what to do about it.

The spectacle of leading scientists falling out about the logic of analysis that underpins almost all of science appears rather alarming. Surely, science has to be faultless, free from error, correct and true. The key point to keep in mind (and one that some contributors to the debate lose sight of at times) is that inference deals in *uncertainty* and probabilistic conclusions, since that is the nature of the evidence science works with. Uncertainty is unsatisfying, especially where decisions and action must rest upon it. Inference is ultimately about quantifying and reducing that uncertainty as far as practically possible, but by definition, it can never eliminate it. This means that rather than there being a unique 'best' solution to a question, usually the best we can do is balance probabilities. It is not then surprising that there is legitimate disagreement about the best way of doing so. In order to consider this debate, let's first review what we've learned about statistical inference in the previous chapters of this book. I find it useful to think of inference as comprising ten stages of work.

Ten steps of inference

1   The formulation of a research question and the identification of appropriate empirical evidence to answer it.

Research questions have diverse origins. Many arise as a result of previous research, since it is unusual for any enquiry to reach a clear and definitive conclusion. Most should bear upon existing research and knowledge, either to test it out or develop it in some way. Robust knowledge is more likely to emerge from the gradual development of a research area and programme across many teams and scientists rather than some bright and original idea that challenges the entire field. This means that a literature review is an essential starting point. This also allows the researcher to see what measurement instruments, if any, have been tested and validated by others working in the field. Such an approach makes possible the accumulation of knowledge over time, even if such a process is uneven and 'dead ends' force the process backwards until a different path is found.

Data sources also have to be considered. There is a growing range of repeated cross-sectional and longitudinal studies in many countries that offer an unprecedentedly wide spectrum of data that is now more accessible than ever before, thanks to the internet and major data archives. However, every data source, no matter how valuable, is limited. Measuring any aspect of social, political or economic life is a complex affair, so that often the specific research questions that can be pursued are limited by the nature of the data available. This means that theory and data have to be considered together. Keep in mind that no amount of expert statistical inference can correct measurement error and that most measurements of social behaviour and consciousness are crude at best.

2   The research design and assumptions.

These are both practical and statistical. The three most important components will be the nature of randomisation, the definition of the target population and protocols for measurement.

Randomisation will often be achieved by sampling, and robust data sourced from a data archive will usually have been based on a thorough sampling design. Typically, however, samples are stratified and clustered, so that one of the first tasks is to understand and correctly apply the weights used in the data, and the implications of these weights for the size of standard errors calculated. Alternatively, experimental designs need to ensure randomisation in the allocation of subjects to treatments and consider any limitations upon that imposed by ethical concerns, for example. Good measurement often also requires blinding, while analysis may have to consider dependencies between samples. Many experimental designs have been developed to maximise the power that may be available with a limited number of experimental subjects, such as crossover or step-wedge designs.

What/who is the target population? Imagine a simple case. We have a social attitude survey of a random sample of adults in a country. First think of the selection effects that arise from the practical organisation of the survey. A comprehensive sampling frame with contact details probably does not exist. If it does, it will inevitably be out of date, reducing the probability that those who move residence more often will be missed. A surrogate frame, such as a list of addresses or phone numbers, will have its own biases. The homeless, those in institutions such as prisons, hotels, hospitals, elder care facilities, student residences or army barracks, may be missed, or those recently installed in new housing developments. Some respondents will be easier to contact than others, easier to persuade to participate than others or respond more truthfully and conscientiously to the questions than others. Finally, having overcome these obstacles, we face a further one. The characteristics of the target population at the time of the survey fieldwork may be of interest, but our real concern may well be to generalise to *future* populations of that kind. Research takes time. Will the characteristics of the target population be the same once the research has been undertaken and published?

Finally, possibly the most important element of design is the measurement process that generates the data. Most people think of measurement in terms of something like a school ruler. You place it next to the object and read off the length. The problem is that most phenomena the social sciences wish to measure are invisible, intangible or subjective. They have to be inferred from their association with other features that are more visible, including aspects of respondents' behaviour or, still more fraught with error, asking respondents to undertake a lightning introspection of their feelings or attitudes and revealing them to us. Too many social scientists and survey designers have a touching faith in respondents' capacity for such self-analysis. Most measurement is expensive and difficult to do well. There is almost always a trade-off between reliability (systems that generate similar results in similar circumstances) and validity (capturing the true phenomenon under investigation). Real data is messy. It inevitably contains errors. Most processes of measurement are far from perfect. This is especially true in the social sciences where many of the core concepts in everyday use (class, status, ethnicity, income, power, identity and religion) typically defy simple or accurate definition. Typically, statistical analyses make assumptions about the nature of the data that is to be analysed, including about its distribution and random selection from a single population. Often, data and its analysis will be robust even to gross violations of these assumptions, but this is not always the case.

3   The choice of appropriate statistical model.

A model is the mathematical representation of the social feature, institution or process that the research attempts to describe or explain. It may be as simple as a summary statistic such as a mean or interquartile range for a single variable, or it might be a regression equation using several variables. The key issue is how well the model simplifies the social characteristics being represented. There is always a trade-off between model parsimony and capturing the complexity and messiness of social processes. The key questions are usually whether the right variables are in the model and whether any violations of assumptions the model makes are likely to be problematic. For example, regression models may assume that a variable is distributed approximately normally but still perform well if in fact the variable distribution departs from this.

In this book, we have tended to focus on very simple models for continuous and dichotomous variables, to keep the mathematics clear. However, variables come with different levels of measurement, so that a range of models and tests have been developed to deal with different circumstances. Frey (2015) is an excellent short guide to the main tests that have been developed to deal with different configurations of evidence.

Managing and preparing data to fit a model, and meticulously documenting how this is done, are usually time-consuming and need to be planned for. Often, it accounts for as much as 90% of the time needed for a project. Such data 'wrangling' can be tedious, but doing it in such a way that the treatment of the data is transparent, and can be reproduced by others, is fundamental to maintaining scientific integrity.

4   The specification of a null hypothesis about the target population.

The null hypothesis often, but not always, takes the form of 'no effect', 'no difference' or predicting that some quantity will equal zero. In some versions of NHST, an alternative hypothesis may also be specified, or it may be logically implied by the rejection of the null if it is a dividing hypothesis. The null hypothesis *always* defines the value of some *population parameter*, since all inference is about translating what we know *with certainty* about our sample data, into probabilistic statements about the population from which our data was taken. The null hypothesis might concern the mean value of a variable, or the proportion of members of the population which take a given value of a variable, either in absolute terms or in relation to another variable. The word *specify* is fundamental. Take the following null hypotheses, for example:

a   Mean earnings of women equal those of men.
b   Graduates were more likely than others to vote Democrat in the last presidential election in the USA.
c   Less than 10% of NGOs in the USA in 2015 had women chief executives.

If you consider them, it soon becomes clear that the target population and parameters they might refer to are poorly specified. For hypothesis (a) do we want to generalise to all men and women globally? Forever? Regardless of age? Including those not currently earning? What constitutes earnings? Wages? Other income? Income from capital, inheritance or other sources? Over what period? Across a lifetime? Last week? We might ultimately want to make a statement about sex and material inequality globally, but to do so with evidence, we need to have a crystal-clear definition of the data we intend to collect. And how equal is equal? Would a difference of one or two percentage points be important? Depending on the context, it might either be vital proof of discrimination or evidence of equality. Hypothesis (b) is more tightly specified, but still needs clarity on how graduate is to be defined, and 'more likely' could be understood in two distinct ways. It could be interpreted as a simple comparison between the proportion of graduates and others who voted Democrat. However, it could also be interpreted as what impact being a graduate had, once other factors, such as gender, age, area of residence, social class, religion and so on, had been controlled for in a regression model. Hypothesis (c) would also need good definitions of 'NGO', 'in the USA' and 'chief executive'. Do we mean, for example, NGOs that are based in the USA, or NGOs that operate there, even if their head office (and chief executive) might lie elsewhere?

These distinctions might appear to be splitting hairs. Discussion of most topics proceeds fluently enough in the literature without constantly defining terms in great detail, but this is absolutely not the case with measurement and data analysis. No inferential statistics can rescue data that is poor because it has been inadequately specified. The immediate inferences that we can draw from data depend upon a careful specification of both the target population and the measurements to be taken from it. *After* that has been successfully accomplished, it is then possible to make a judgement about how far beyond that specific measurement or target population it may be safe to generalise by claiming that it is typical of other measurements that could have been made, or other target populations that might have been measured.

You will find a wide range of wording to describe the conditional description that the null hypothesis is true. There are two principal types of null hypotheses. 'Dividing' hypotheses always imply that the null can be rejected in only one 'direction'. Examples are (b) and (c) above. Rejecting a null hypothesis that graduates were *more* likely to vote Democrat must imply that their probability of voting Democrat was equal to or less than non-graduates. The proportion of NGOs is either less than 10% or it is not. In other circumstances, we may have no prior expectation of the direction in which an association might lie. Perhaps men's earnings are greater than women's, perhaps they are lower. In such a circumstance, we do not know in which tail of the sampling distribution our test statistic might lie in the case of its rejection.

As shorthand, instead of writing 'assuming the null hypothesis were true', 'under the condition that the null hypothesis is correct', 'given the condition that the null hypothesis holds', 'conditional on the null hypothesis' or 'conditioning under the null hypothesis', we can just write *under the null*.

5   Calculation of a test statistic.

The *test statistic* describes the distance between the parameter value expected under the null and that observed in the sample data. The larger the absolute value of the test statistic, the greater the inconsistency between the data observed and that expected under the null. Test statistics, such as z, t, F and x^2 have a known sampling distribution under the null. Because of this, the distance between the observed test statistic and that expected under the null can be expressed in *standard errors* (standard deviations of a sampling distribution), and the area within the sampling distribution lying beyond the observed value of the test statistic can always be calculated.

6   Calculation of a *p*-value.

The *sampling distribution* under the null, and the location of the test statistic within it, gives the probability of obtaining the test statistic observed, or one more extreme, under the null. This is usually referred to as the *p*-value. Typically, the test statistic lies towards the tail of the sampling distribution. These distributions have different shapes. Some are symmetric, so that the test statistic might lie in either tail. This raises the possibility of *one-tailed* and *two-tailed* tests. Where a test is two tailed, the probability of observing any given value of the test statistic under the null will be higher.

7   A conclusion about the model.

No model is perfect, and all models are 'wrong' in some way. Their job is to simplify the complexity and messiness of social relations, strip away superfluous variation and reveal underlying patterns, structures, associations or causal paths. There will be no single 'best' model to do this. As well as interpreting the substantive output from the model, assessing the performance of a model will involve diagnostics: the examination of residuals for patterns, consideration of outlying values and so on. Although some attention will be paid to whether or not any result is 'significant', on its own, this may not be a decisive consideration. As we have seen, large samples often throw up trivial but significant results. Where data is sparse, an association that is compatible with accepting the null hypothesis may still be important evidence.

8   A conclusion about the target population.

A null hypothesis may be rejected on the grounds that the probability of observing the data obtained is so small that this constitutes *provisional* evidence that the null does not hold. *Under some conditions*, the implications of the rejection of the null are clear. For example, if we reject a null that 'women's mean earnings are higher than or equal to men's', it logically implies evidence that they are lower. Other times, the rejection of the null simply indicates that some hypothesised process is at work, but further research is needed to establish what form or magnitude it might take. A null hypothesis that is *not* rejected may be 'accepted' but is *not* proved. Being unable to rule out 'no effect' or 'no difference' is not the same as producing good evidence that there is indeed no effect or no difference. It could be that our test was simply not powerful enough to detect an effect that was in fact there.

Any inferential conclusion is inherently risky. It weighs up evidence that is never conclusive and decides upon a balance of probabilities for and against the null.

Thus, any conclusion may be either correct (rejecting a false null hypothesis or accepting a true one) or commit a Type I or Type II error:

- A Type I error occurs when a true null hypothesis is rejected. A Type I error is a false positive: the conclusion that some effect, difference or result is present in the target population, when in fact it is not. The *p*-value is equal to the probability of making Type I errors.
- A Type II error occurs when a false null hypothesis is accepted. A Type II error is a false negative: the conclusion that some effect, difference or result is not present in the target population, when in fact it is. It is more difficult to estimate the probability of making a Type II error since it depends upon the size of the effect that is missed.

Keep in mind too that NHST gives us the probability of our data under the null, not the probability of the truth or otherwise of our hypothesis. Similarly, confidence intervals describe what we can expect from our procedure, and our level of confidence in it, *not* the probability that any individual result or population parameter that we estimate falls within the interval.

9   A substantive conclusion.

Together with its implications for future research, this is the whole point of inference, so it is hardly surprising that this part of the process has attracted the most discussion and debate. There is not too much to argue about in the mathematics of probability, sampling or distributions. No one, for example, argues about the formula for defining a Gaussian distribution or the nature of its shape. But there is ample opportunity for discussing many other issues. This is not because the scientific logic has not been sorted out, but rather because the implications of that logic, for how results ought to be interpreted in widely differing contexts are not always clear. In well-designed empirical research, with appropriate data, good model and carefully specified null, the implications of the research may indeed be very clear: at least to the authors of the research. Then, it is the job of the rest of the scientific community to explore alternative accounts of the results, weaknesses in the operationalisation of concepts or choice of a model or any other aspect of the research.

10   Reproduction and replication.

This is substantively the most important, but in many disciplines, most neglected aspect of inference. Even the strongest inferential conclusion, with a tiny *p*-value, derived from a robust and appropriate model, using data of the highest quality, that adequately considers both variation and precision; taken from a genuinely random sample from a clear target population and appropriately cautious about generalising beyond it, is the *beginning* rather than the end of *scientific* inference. One reason to treat it as only the start of a research programme is simple human error. There are hundreds of steps from the initial formulation of a research question to the results of an inferential test or data exploration. One error in any of them, by anyone, might undermine everything. Thus, the first check that ought to be made by the scientific community is to *reproduce* the analysis undertaken. This requires sufficient transparency and full and robust reporting of the original analysis to enable this to be done efficiently and effectively. This also makes meta-analysis possible. The second check is to keep in mind the dark side of randomness. Given the industrial volume of modern science, it is always possible that the beautifully fitting model and elegant explanations that it is consistent with come from a point far into the tail of the sampling distribution under the null. It is therefore best treated, like Fisher's agricultural tests, as a *signal* that warrants further investigation. Again, the only antidote to this is *replication*, using other data sources, other ways of measuring the phenomenon under investigation and perhaps other models.

The logic of hypothesis testing has usually been set out imagining a single artisan scientist or team of scientists formulating a null hypothesis, deciding what level of evidence against the null they require to reject it, collecting relevant data and then testing the hypothesis on it. The prior specification of this hypothesis is an essential part of this logic, because any set of data, once collected, may yield evidence to reject thousands of potential null hypotheses. Given the probabilistic nature of evidence against the null, we could expect a very large proportion of these rejected nulls to constitute little more than random noise in the data. Thus, trawling the data in search of nulls to reject is outlawed as 'data snooping' or '*p*-hacking'. A result 'discovered' by such methods by definition has a probability of producing the data observed under the null of exactly one. Yet, in practice, this is neither how science proceeds nor how it ought to proceed. Most scientists spend most of their time exploring data and learning from it. It would be utterly counterproductive to 'outlaw' such activity to protect the statistics of inference. However, it does mean that research that has sound inference as its objective must be separated in some way from data exploration. Where data is cheap and plentiful, this is not difficult. The reverse is true when it is sparse and expensive to collect.

In the original artisan model, a 'significance level' of 5%, or sometimes 1% would be set for the probability of observing the data under the null in order to reject it. Such a rule had two great benefits. It massively simplified calculation in an era before information technology when a 'computer' was one of several dozen workers in a room (usually women) doing endless hand calculations. It also set a standard with a useful ability to compare apples and bicycles. Scientists might concentrate their research on results that provisionally looked 'significant'. This has encouraged what are sometimes called 'bright lines' across the entire institutional organisation of scientific work, from university departments to government institutes, pharmaceutical, medical or innovatory technology enterprises to publishers of scientific journals and their editorial boards and reviewers, where a p-value of 0.049 is 'a result' and worthy of publication and one of 0.051 is an unevent which gets consigned to a dusty bottom drawer. This is not only logical nonsense, it may encourage all manner of perverse impacts. If there are multiple scientists and teams in any substantive research area, working with commonly available data, curated in data archives and distributed over the internet, it is highly likely that most of the thousands of rejectable null hypotheses lurking in that data will be unearthed. How is this different from the 'data snooping' of a rogue scientist?

On the other hand, it is too easily overlooked that some kind of bright line is inevitable in *any* situation where decisions have to be made. Having them, no matter how perverse they may seem to be in individual cases, is almost certainly preferable to leaving everything to subjective judgement. An excellent analogy with p-values is student exam results. Often there is a 'bright line' of 50%. The hapless student who scores 49.4% fails. The lucky one who scrapes 49.5% passes (if marks are rounded). Only a fool would conclude that there was any material difference between the performance of the two students, but that cannot, and should not, mean that the bright line should be abandoned. The only alternative would be subjective judgement by someone given authority to make it, with all the inevitable biases, abuses, suspicions and rancour that it would involve.

The six principles of the ASA

Concerned about the abuse of significance tests and p-values in scientific research, the ASA agreed on six principles to guide statistical work in empirical research. It is worth considering each of them in relation to what we've covered in this book.

1 *p-Values can indicate how incompatible the data are with a specified statistical model*: A p-value provides one approach to summarising the incompatibility between a particular set of data and a proposed model for the data The smaller the p-value, the greater the statistical incompatibility of the data with the null hypothesis, if the underlying assumptions used to calculate the p-value hold. This incompatibility can be interpreted as casting doubt on or providing evidence against the null hypothesis or the underlying assumptions.

If you have understood this book, this principle will appear self-evident. The principle emphasises that any null hypothesis *also* implies a specific statistical model.

This model not only contains *assumptions*, which must be appropriate, but its incompatibility with a *particular* set of data *can* also provide evidence against a null. None of this is automatic. The data might be contaminated or unusual. It might be incompatible with the model because the latter has been ill thought out. There is always room to think carefully about how far that model is likely to be an adequate one, and how far it over-simplifies a messy reality in order to keep the model tractable, as Tukey and the exploratory data analysis movement made clear. No model is perfect, any model is a compromise, and all models are (ultimately) wrong, so that some way to reject a null hypothesis can always be found. The challenge is to propose a null that makes substantive sense, so that the interpretation resting on its acceptance or rejection is clear, and to be aware of alternative models that might be specified and to focus on models that are likely to prove useful and relevant.

2   *p-Values do not measure the probability that the studied hypothesis is true, or the probability that the data were produced by random chance alone*: The *p*-value is neither. It is a statement about data in relation to a specified hypothetical explanation and is not a statement about the explanation itself.

As has been stressed throughout this book, $P(\text{data} \mid H_0) \neq P(H_0 \mid \text{data})$. The parsing of the second sentence might raise an eyebrow or two however. In what way is a statement about data *in relation to* a specified hypothetical explanation *not* about 'the explanation itself'? It could indeed be argued that the logic of $P(\text{data} \mid H_0)$ means that it is the data that is the object of the analysis and not the null hypothesis, but this seems to me to be an exercise in hair splitting. It is the capacity of the data to provide evidence for or against the null that gives the data any interest for us in the first place! However, there is another fundamental sense in which this distinction is vital. If we focus on the probability of the data, that stops us from jumping to unsound conclusions about the hypothesis that do not take the nature of the data into account. Thus, if a small underpowered study fails to reject a null, that tells us more about the data than whether the null hypothesis is true. Conversely, a 'finding' of unimportant consequence made possible by a large sample of data may be used to reject a null and even point towards an alternative hypothesis, but if it is of no substantive consequence, who cares!

3   *Scientific conclusions and business or policy decisions should not be based only on whether a p-value passes a specific threshold*: Practices that reduce data analysis or scientific inference to mechanical 'bright-line' rules (e.g. '$p < 0.05$') for justifying scientific claims or conclusions can lead to erroneous beliefs and poor decision-making. A conclusion does not immediately become 'true' on one side of the divide and 'false' on the other. Researchers should bring many contextual factors into play to derive scientific inferences, including the design of a study, the quality of the measurements, the external evidence

for the phenomenon under study and the validity of assumptions that underlie the data analysis. Pragmatic considerations often require binary 'yes–no' decisions, but this does not mean that p-values alone can ensure that a decision is correct or incorrect. The widespread use of 'statistical significance' (generally interpreted as '$p \le 0.05$') as a license for making a claim of a scientific finding (or implied truth) leads to considerable distortion of the scientific process.

This is a key paragraph. The main principle ought to be clear if you followed the distinction between scientific and statistical inference in Chapter 1. However, the words *only* in the first sentence and *alone* in the penultimate one are the key terms whose strength and interpretation is argued about vigorously. It is perfectly logical to insist that p-values alone are an insufficient metric. A poorly designed study that yields a low p-value does not become either scientific or useful by virtue of throwing up a finding that would be classified as 'significant'. This argument is something of a straw person. Journal editors or civil servants responsible for food or drug testing and safety are not usually naive about study design, nor likely to give credibility to any study solely by virtue of its p-value. What *is* true is that policymakers, journalists and others may be taken in much more easily.

The replication crisis is *within* science, however. Imagine a world in which, because of the desire to escape binary decisions made on small margins, the 'bright lines' were to be abandoned. It is indeed possible that some better decisions might be made. It is surely equally possible that some altogether worse ones might be made too. Findings of well-conducted studies with $p = 0.49$ that might previously have been published might be shelved for quite other reasons. Scientists may be no better than others at allowing unconscious biases and expectations colour their attitudes towards a piece of research in addition to any assessment about the quality of its overall design. History is littered with ideas which were lampooned and ridiculed when first advanced, only to become accepted later. The 'Big Bang' was named thus to discredit what was initially assumed to be a risible theory. Should drug companies convinced of the value (and profitability) of their new product be given the latitude to convince or cajole the regulatory authorities into letting $p = 0.51$ stand?

4   *Proper inference requires full reporting and transparency*: Cherry picking promising findings, also known by such terms as *data dredging, significance chasing, significance questing, selective inference* and '*p-hacking*', leads to a spurious excess of statistically significant results in the published literature and should be vigorously avoided. Researchers should disclose the number of hypotheses explored during the study, all data collection decisions, all statistical analyses conducted and all p-values computed. Valid scientific conclusions based on p-values and related statistics cannot be drawn without at least knowing how many and which analyses were conducted, and how those analyses (including p-values) were selected for reporting.

This logic is impeccable, and in some areas of medical research, it is now in effect what happens, with the preregistration of protocols before research begins. Indeed, robust inferential research should always be conducted this way. However, such explicitly inferential research is not the only way, as we have seen, of arriving at practical inferential conclusions. Often, we are in the business of drawing more informal inferences from the robust exploration and description of data. The organisation and conduct of exploratory data analysis runs absolutely counter to the guidance here. The key issue, not really addressed, is how best to marry and coordinate the two. The solution here probably lies less with protocols for multiple potential comparisons than with considering the organisation of science beyond individual scientists or research groups. I discuss this further below.

5   *A p-value, or statistical significance, does not measure the size of an effect or the importance of a result*: Statistical significance is not equivalent to scientific, human or economic significance. Smaller *p*-values do not necessarily imply the presence of larger or more important effects, and larger *p*-values do not imply a lack of importance or even a lack of effect. Any effect, no matter how tiny, can produce a small *p*-value if the sample size or measurement precision is high enough, and large effects may produce unimpressive *p*-values if the sample size is small or measurements are imprecise.

This is also a point we have made many times in the course of this book and is utterly uncontroversial. However, one could also have argued, given similar sample sizes and measurement precision, that a *p*-value can be an *excellent* way of quickly discriminating between associations that *may* prove worthy of further investigation. In the conditions Fisher faced at Rothamsted that we described in Chapter 3, for example, exactly such a procedure brought order and insight to what had been, in retrospect, the amassing of data with little ability to make any important sense of it. Less uncontroversial is the final principle in the statement:

6   *By itself, a p-value does not provide a good measure of evidence regarding a model or hypothesis*: Researchers should recognise that a *p*-value without context or other evidence provides limited information. For example, a *p*-value near 0.05 taken by itself offers only weak evidence against the null hypothesis. Likewise, a relatively large *p*-value does not imply evidence in favour of the null hypothesis; many other hypotheses may be equally or more consistent with the observed data. For these reasons, data analysis should not end with the calculation of a *p*-value when other approaches are appropriate and feasible.

From the six principles to ATOM

It is best to consider this final paragraph alongside the development of the debate after 2016. In March 2019, the ASA published a special edition of its journal devoted to inference, bringing together papers from the symposium it had organised the previous year, and introduced by an editorial comment (Wasserstein et al., 2019) which went a step further than the originally agreed 'six principles' and suggested a new mnemonic 'ATOM':

> We conclude, based on our review of the articles in this special issue and the broader literature, that it is time to stop using the term statistically significant entirely. Nor should variants such as significantly different, $p < 0.05$ and nonsignificant survive, whether expressed in words, by asterisks in a table, or in some other way.
>
> **A**ccepting uncertainty as inevitable is a natural antidote to the seductive certainty falsely promised by statistical significance.
>
> **T**houghtful research prioritises sound data production by putting energy into the careful planning, design, and execution of the study ... considers the scientific context and prior evidence ... looks ahead to prospective outcomes in the context of theory and previous research.
>
> Be open to 'open science' practices ... such as public pre-registration of methods, transparency and completeness in reporting, shared data and code, and even pre-registered ('results-blind') review. Completeness in reporting, for example, requires not only describing all analyses performed but also presenting all findings obtained, without regard to statistical significance or any such criterion. **O**penness also includes understanding and accepting the role of expert judgment. ... 'Judgment is necessarily subjective, but should be made as carefully, as objectively, and as scientifically as possible'.
>
> Be **M**odest about the role of statistical inference in scientific inference. 'Scientific inference is a far broader concept than statistical inference'.

In March 2019, more than 800 scientists signed a letter to *Nature*, calling for *p*-value thresholds to be abandoned entirely. They argued (Amrhein et al., 2019):

> Let's be clear about what must stop: we should never conclude there is 'no difference' or 'no association' just because a P value is larger than a threshold such as 0.05 or, equivalently, because a confidence interval includes zero. Neither should we conclude that two studies conflict because one had a statistically significant result and the other did not. These errors waste research efforts and misinform policy decisions. We ... call for the entire concept of statistical significance to be abandoned. We are not calling for a ban on P values. Rather, and in line with many others over the decades, we are calling for a stop to the use of P values in the conventional, dichotomous way – to decide whether a result refutes or supports a scientific hypothesis ... we are not advocating a ban on P values, confidence intervals or other statistical measures – only that we should not treat them categorically.
>
> We must learn to embrace uncertainty. One practical way to do so is to rename confidence intervals as 'compatibility intervals' and interpret them in a way that avoids overconfidence. Specifically, we recommend that authors describe the practical implications of all values inside the interval, especially the observed effect (or point estimate) and the limits .... just because the interval gives the values most compatible with the data, given the assumptions, it doesn't mean values outside it are incompatible; they are just less compatible ... like the 0.05 threshold from which it came, the default 95% used to compute intervals is itself an arbitrary convention.

Some found the ASA's original six principles too finely balanced and modest and pushed for the more decisive rejection of the language of significance – 'bright-line rules' such as p-values above or below 5% and confidence intervals of 95% – without further specification of the range of substantive conclusions they might be compatible with. I think there were many good reasons why the ASA stopped where it did in 2016, and I very much doubt that the suggestions of Wasserstein et al. will be taken up in the way they hope, mostly because by trying to go further, it may be that they have only managed to confuse the issue. The call to 'abandon significance' is made alongside a large number of quite uncontroversial statements about good research practice that I think any conscientious scientist would endorse. The problem is that few of them bear directly on the key issue of the relative importance of significance, its proper reporting and interpretation and what purposes it can and cannot serve, nor on whether dichotomous decision-making based on p-values ought to be abandoned.

Both Wasserstein et al. (2019) and Amrhein et al. (2019) would have done well to think more carefully about their language. Calling for the 'entire concept of statistical significance to be abandoned' is neither what they in fact do nor would it make any sense. Who wants to abandon the concept of data being incompatible with a null hypothesis? It is the *only* calculation we can reliably make about sample data without making other prior assumptions. Having made that calculation, embracing uncertainty is not enough. If we are not ultimately interested in whether or not the null is true, what point would the research have had in the first place? This is rhetoric, not statistics. It is true that making binary decisions solely on the basis of a p-value is stupid. But such an argument is a straw person. Most journal editors would insist on a sufficiently robust research design to ensure that the p-value obtained was meaningful.

Conversely, embracing uncertainty is laudable, but it has its limits. One of the objectives of statistical literacy is, and ought to continue to be, to educate everyone in the difficulties and expense of good measurement, the inevitable uncertainty that surrounds *all* our estimates about economy and society and how that uncertainty broadens when we attempt to make inference about the future rather than simply record the past. One of the paradoxes of the era of 'big data' is that citizens, managers, policymakers and almost anyone who is neither a statistician nor a scientist all tend to assume that 'the data is out there' when it very rarely is. Were they more aware of the vicissitudes of measurement, they might expect less certainty.

By contrast, anyone faced with making a decision has to have a criterion to make it by. Wasserstein (2019), Amrhein (2019) and colleagues concede this in the case of routine quality management. They *must* do so, because it would be nonsense not to. This begs the question, why is what is good enough for routine quality management not sufficient for science? One of the many questions that science faces are all kinds of decisions about priorities. Those decisions require some common currency within which calculations and decisions can be made across widely different substantive research areas or problems. Significance is a method for literally signalling a promising line of enquiry. As such, it is a useful currency that it would make little sense to abandon. This is still more the case if there is not an alternative that commands widespread support and is well understood.

This raises a further problem. Comparing the misuse of a widely used statistical procedure such as NHST with the proper, cautious and expert use of alternative approaches is illogical. Had the scientific community embraced another set of procedures as standard, it is likely that they too would be misused. In a perfect world, every scientist would also be a good statistician, or every research team would have one. In an imperfect one, calling out poor statistical practice may not be evidence of widespread abuse but rather evidence of a robust response to it. This takes us to a further consideration that actually coincides well with some of the other conclusions reached by those calling for a significance 'ban'.

Contemporary scientific research

Neither Fisher, Pearson and Neyman nor any of the other pioneers of the statistics of inference could probably have imagined the change in scale of the scientific community that has taken place over the last century. It is hard to get good estimates of the volume of scientific research. A reasonable proxy is probably the volume of publication of scientific papers, but not all research ends up in such papers, and some papers, and the journals that publish them, are bogus vanity efforts, fuelled, amongst other things, by the rise of poorly designed metrics and league tables in higher education and the 'pressure to publish' they create, especially on younger academics. The current growth rate of journal articles is around 5% per year. There are more than 30,000 scientific journals and another 10,000 or so bogus ones. These journals publish more than 2.5 million articles a year, and rising, and some estimates of the number of bogus articles within this total put it as high as 0.4 million.

When scientific activity reaches this sort of scale, the artisan model of scientific production that lies behind Fisher's approach becomes a less good guide to discriminating between potentially interesting results and those that can be safely discarded. If one researcher deals with one piece of data, formulating a single hypothesis beforehand, and then calculates the probability that a null hypothesis is true, a 5% threshold might make sense as a balance between discarding what might prove to be an interesting result and following up what might prove to be a red herring.

Now, imagine that exactly the same data is now examined by 20, 200 or 2000 research teams, again each formulating their own distinct hypotheses. If the various hypotheses are statistically independent of each other, on average, 1, 10 or 100 researchers will be declaring random noise to be a genuine research finding. However, such a scenario is unlikely. If the researchers have diligently read the available literature on a research question, it would be more likely that their hypotheses were associated in some way, increasing the chance of some types of noise passing through significance filters. Imagine a slightly different scenario. This time each researcher has their own *data*, all of which we could treat as a sample drawn from all the data they *could* have collected. In neither scenario is there any *p*-hacking, but the end result is much the same.

This is not a good way to do science, but the solution does not lie in perfecting inference tools alone. Any inference procedures must balance the risk of false positives against that of false negatives. The problem is rather that these may no longer be probabilities that any individual scientist working on a research problem or using a data set that is also used by others can possibly calculate.

There are two other challenges that address the issues raised in the debate over inference more directly and may do more to resolve them. Both require better communication between statisticians and scientists. The first is what might be called (with apologies to Max Weber) traditional scientific statistical social action. The logic of statistical inference is hard work. It is a skill that takes time and experience to master. It is easy to make errors. One of the results of this is that disciplines tend to fall back on favoured methods of statistical analysis whose main rationale is habit rather than a comprehensive review of what might best fit the purpose pursued. The apprenticeship model of learning from others reinforces this trend. Each subject tends to have its favourite statistical tests, measures of association and ways of reporting results. Were there a more energetic exchange between statistics as a university discipline and its application in other scientific disciplines, such behaviour would become harder to sustain.

The other challenge is reproducibility, replication and meta-analysis. It makes little sense for individual scientists to continue behaving like cottage artisans in the era of industrial scale science. The ultimate solution to the problem of multiple comparisons, underpowered studies and even bogus science in vanity journals is to take much greater advantage of the power of data management and processing technology to archive and curate the *results* of data analysis rather than only the original raw data. Tentative steps in this direction have been made in many disciplines, and it is surely where the statistical future lies, bringing the power of data science to bear on the work of individual research teams such that their results become part of a more comprehensive evidence landscape. One of the essential tools in such an approach will continue to be inference.

Further Reading

The ASA statement (Wasserstein, 2016) and accompanying articles, together with the follow-up (Wasserstein et al., 2019), are the best place to follow current thinking on how best to manage sound inference in academic research. Ziliak (2016) is clear headed and helpfully thinks about the whole process of research. The article by Ioannidis (2005) that kicked off the current discussion about the misuse of inference rules is still worth reading. It is noteworthy that Ioannidis himself argues that abandoning p-values would do more harm than good (Ioannidis, 2019).

GLOSSARY

Bernoulli trial: A trial with only two outcomes.

Binomial coefficient: The number of combinations of k items that can be chosen from a set of n items, for example, the number of ways of obtaining $k = 4$ 'heads' from $n = 10$ flips of a coin.

Categorical data/variables: Variables which take the form of discrete categories rather than a continuous value range – e.g. the level of formal education qualification held by a respondent.

Central limit theorem: Repeated random samples taken from a population distribution with a given mean and variance, regardless of the shape of this distribution, will produce a sampling distribution which is normally distributed around the mean of the population distribution. The standard deviation of the sampling distribution will vary with the square root of the sample size and variance of the population distribution.

Chi-square (χ^2): With k degrees of freedom is the sum of square of k standard normal random variables.

Confidence: The probability in the long run of a sampling procedure producing a correct result.

Confidence interval: Given a level of confidence, the range of values within which a population parameter is estimated to lie.

Continuous data/variables: Variables at the ratio or interval levels of measurement take values that may only be described by the interval within which they fall.

Correlation: The degree of association between two variables, usually expressed on a scale from zero to one.

Critical region: The range of values for a sample statistic corresponding to the rejection of the null hypothesis at a chosen probability level.

Dummy variable: A variable taking only two values, often used in regression and conventionally taking the values zero and one.

Histogram: A graphic that displays a continuous variable, in which the area corresponding to a range of values displayed on the horizontal axis equals the proportion of all observations in that value range.

Hypothesis: A statement about a population which inference attempts to refute or confirm.

Inference: The process of making statements about or descriptions of a target population based on analysis of a random sample drawn from it.

Intercept: In the geometric representation of a regression equation, the point at which the regression line crosses the vertical axis, or the value of $\hat{Y}$ for $X = 0$.

Law of large numbers: The average results of any independent trial repeated a large number of times will converge toward a limit which equals the probability of obtaining that result.

Mean: The arithmetic average of a group of values.

Missing Data: Cases in a data set where no value has been recorded for a particular variable, either because of some failure in the collection process or because no such value exists.

NHST: Inference based on testing null hypotheses at a chosen significance level.

Normal distribution: A symmetrical distribution in which the mode, median and mean value coincide, and the proportion of observations declines as their distance from the mean increases.

Null hypothesis: A statement about a population for which a sampling distribution can be constructed, often of the form that some value equals zero, or that there is no difference between two groups defined by a categorical variable.

Operationalisation: The translation of an abstract concept into an empirical variable or set of variables that can be measured in some way.

Parameter: A characteristic of a population – e.g. proportion, mean, standard deviation and so on.

Point estimate: An estimate of the value of a population parameter taken from a random sample.

Population: A collection of people, objects or anything else whose measurement is of interest to us.

Power: The long run probability of a test or analogous procedure to reject a null hypothesis that is in fact false.

Probability: The proportion of times a result of interest occurs in an identical independent trial repeated many times; the rational degree of belief, given the available evidence, in the truth of some hypothesis.

p-Value: The probability of obtaining some result or measurement under the assumption that the null hypothesis is true for the population.

Random sample: A sample in which every member of a population has a known probability of being selected.

Regression: The use of one or more variables to produce an equation estimating the value of another variable.

Reliability: The extent to which a process (e.g. measurement or an experimental procedure) can be repeated and yield the same result.

Sample space: The set of outcomes from any trial.

Sampling distribution: The distribution of the estimate for a population parameter produced by all random samples of size N drawn from a population.

Sigma: σ, Greek letter and symbol for the standard deviation.

Slope: The steepness of a regression line, representing the average change in $\hat{Y}$ for a unit change in X.

Standard deviation: A measure of the dispersion of a range of values, in which a higher value indicates more dispersion.

Standard error: The standard deviation of a sampling distribution.

Sum of squares: The sum of squared residuals in a distribution to the mean of that distribution or to predicted value for it.

Trial: A set-up in which an outcome from some process is not uniquely determined.

Type I error: To *reject* a null hypothesis that is in fact true, sometimes referred to as a 'false positive' result.

Type II error: To *accept* a null hypothesis that is in fact false, sometimes referred to as a 'false negative' result.

Variance: The average of the squared residuals from the mean of a distribution.

Check out the next title in the collection *Survey Research and Sampling*, for guidance on Survey Research.

REFERENCES

Agresti, A., & Franklin, F. (2007). *Statistics*. Pearson.

Allison, P. D. (1999). *Multiple regression: A primer*. Pine Forge Press.

Amrhein, V., Greenland, S., & McShane, B. (2019). Scientists rise up against statistical significance. *Nature, 567*(7748), 305–307. https://doi.org/10.1038/d41586-019-00857-9

Atkins, D. C., Baucom, D. H., & Jacobson, N. S. (2001). Understanding infidelity: Correlates in a national random sample. *Journal of Family Psychology, 15*(4), 735–749. https://doi.org/10.1037/0893-3200.15.4.735

Blastland, M., & Spiegelhalter, D. (2013). *The Norm chronicles*. Profile Books.

Comstock, G. C. (1922). Biographical memoir Benjamin Apthorp Gould 1824–1896. *Biographical Memoirs of National Academy of Sciences, 17*, 153–170.

Connelly, M. (2008). *Fatal misconception*. Harvard University Press.

Dawson, L. (2004). The Salk polio vaccine trial of 1954: Risks, randomization and public involvement in research. *Clinical Trials, 1*(1), 122–130. https://doi.org/10.1191/1740774504cn010xx

Driscoll, P., Lecky, F., & Crosby, M. (2000). An introduction to statistical inference—3. *Emergency Medicine Journal, 17*(5), 357–363. https://doi.org/10.1136/emj.17.5.357

Ehrlich, P. (1968). *The population bomb*. Ballantine Books.

Fisher, R. A. (1925). *Statistical methods for research workers*. Oliver & Boyd.

Freedman, D., Pusani, R., & Purves, R. (1978). *Statistics*. W. W. Norton.

Frey, B. B. (2015). *There's a stat for that!* Sage.

Gelman, A., Carlin, J. B., Stern, H. S., Dunson, D. B., Vehtari, A., & Rubin, D. B. (2013). *Bayesian data analysis* (3rd ed.). Chapman & Hall/CRC. https://doi.org/10.1201/b16018

Gigerenzer, G., & Hoffrage, U. (1995). How to improve Bayesian reasoning without instruction: Frequency formats. *Psychological Review, 102*(4), 684–704. https://doi.org/10.1037/0033-295X.102.4.684

Goodman, S. N. (2016). The next questions: Who, what, when, where and why [Online supplemental material]. *The American Statistician, 70*(2). https://doi.org/10.6084/m9.figshare.3085162.v6

Hacking, I. (1990). *The taming of chance.* Cambridge University Press. https://doi.org/10.1017/CBO9780511819766

Hacking, I. (2001). *An introduction to probability and inductive logic.* Cambridge University Press. https://doi.org/10.1017/CBO9780511801297

Independent Breast Screening Review. (2013). The benefits and harms of breast cancer screening: An independent review. A report jointly commissioned by Cancer Research UK and the Department of Health. *British Journal of Cancer, 108*(11), 2205–2240. https://doi.org/10.1038/bjc.2013.177

Ioannidis, J.P. A. (2005) Why Most Published research findings are false. *PLOS Medicine August 30.* https://doi.org/10.1371/journal.pmed.0020124

Ioannidis, J. P. A. (2019). The importance of predefined rules and prespecified statistical analyses: Do not abandon significance. *JAMA Journal of the American Medical Association, 321*(21), 2067–2068. https://doi.org/10.1001/jama.2019.4582

Jørgensen, K. J., Hróbjartsson, A., & Gøtzsche, P. C. (2009). Divine intervention? A Cochrane review on intercessory prayer gone beyond science and reason. *Journal of Negative Results in Biomedicine, 8,* Article 7. https://doi.org/10.1186/1477-5751-8-7

Kahneman, D. (2011). *Thinking fast and slow.* Penguin.

Kruschke, J. (2014). *Doing Bayesian data analysis* (2nd ed.). Academic Press.

Laplace, P. (1951). *Philosophical essay on probabilities.* Dover.

Marsh, C., & Elliott, J. (2008). *Exploring data* (2nd ed.). Polity Press.

Murray, D. B., & Teare, S. W. (1993). Probability of a tossed coin landing on edge. *Physical Review E, 48*(4), 2547–2552. https://doi.org/10.1103/PhysRevE.48.2547

Newsom, S. W. B. (2003). Pioneers in infection control – Joseph Lister. *Journal of Hospital Infection, 55*(4), 246–253. https://doi.org/10.1016/j.jhin.2003.08.001

Office for National Statistics (2016) Teaching Excellence Framework: Review of Data Sources, Methodology Advisory Service, June.

Raper, S. (2019). Turning points: Fisher's random idea. *Significance, 16*(1), 20–23. https://doi.org/10.1111/j.1740-9713.2019.01230.x

Salsburg, D. (2002). *The lady tasting tea: How statistics revolutionized science in the twentieth century.* Owl Books.

Spiegelhalter, D. (2019). *The art of statistics.* Penguin.

Steiger, J. H. (1990). Structural model evaluation and modification: An interval estimation approach. *Multivariate Behavioral Research, 25*(2), 173–180. https://doi.org/10.1207/s15327906mbr2502_4

Tetlock, P., & Gardner, D. (2015). *Superforecasting: The art and science of prediction.* Random House.

Treiman, D. J. (2009). *Quantitative data analysis.* Wiley.

Walker, H. W. (1940). Degrees of freedom. *Journal of Educational Psychology, 31*(4), 253–269. https://doi.org/10.1037/h0054588

Wasserstein, R. L. (Ed.). (2016). ASA statement on statistical significance and p-values. *The American Statistician, 70*(2), 131–132. https://doi.org/10.1080/00031 305.2016.1154108

Wasserstein, R. L., Schirm, A. L., & Lazar, N. A. (2019). Moving to a world beyond "p < 0.05." *The American Statistician, 73*(1), 1–19. https://doi.org/10.1080/000313 05.2019.1583913

Ziliak, S. T. (2016). Statistical significance and scientific misconduct: Improving the style of the published research paper. *Review of Social Economy, 74*(1), 83–97. https://doi.org/10.1080/00346764.2016.1150730

Ziliak, S. T., & McCloskey, D. (2008). *The cult of statistical significance: How the standard error costs us jobs, justice, and lives.* University of Michigan Press. https://doi.org/10.3998/mpub.186351

INDEX